PRACTICAL COLOR GENETICS FOR LIVESTOCK BREEDERS

T0344886

PRACTICAL COLOR GENETICS FOR LIVESTOCK BREEDERS

D. Phillip Sponenberg

Virginia-Maryland College of Veterinary Medicine, Virginia Polytechnic Institute and State University

Also by D. P. Sponenberg

A Conservation Breeding Handbook
Equine Color Genetics (first through fourth editions)
An Introduction to Heritage Breeds
Managing Breeds for a Secure Future, Strategies for Breeders and Breed Associations.
Second Edition

5m Books

Published by
5M Books Ltd
Lings, Great Easton
Essex CM6 2HH, UK
Tel: +44 (0) 330 1333 580
www.5mbooks.com

A Catalogue record for this book is available from the British Library
ISBN 9781789181357
eISBN 9781789181685
DOI 10.52517.9781789181685

Book layout by Servis Filmsetting Ltd, Stockport, Cheshire
Printed by Ömür Printing&Binding, Istanbul, Turkey
Photos by the author unless otherwise indicated, illustrations by Elaine Leggett

Contents

Preface

This book is the fruit of decades of my interest in color genetics across a broad range of species. My interest began as a child with a project that lasted several years, breeding and raising guinea pigs. Careful records were kept, and after all those years the net profit was one dollar. But the real net gain was an abiding interest in genetics, color, and animal breeding that carried over into adulthood. My interests began to cover an ever-expanding list of species. The first published works involved horses, to which have been added goats, sheep, cattle, alpacas, and dogs.

My work and interest in the color of species of livestock has been inspired by a host of livestock breeders and scientists with whom I have shared an interest in color genetics over decades. Several of these people have become long-term colleagues and friends. Over the years they have stimulated me to think through a host of difficult issues concerning the genetics of livestock color. My life has been enriched both professionally and personally by these people, and I am grateful to each of them for having been a part of my life.

Several scientific colleagues are especially important for contributing to a detailed and accurate understanding of coat color genetics. The crew at COGNOSAG (Committee on Genetic Nomenclature of Sheep and Goats) stands out as one unifying group for early progress in the color genetics of these two species. In the 1980s we would get together once a year at Les Doux Moulins in Provence, southern France. We would talk genetics during the day, eat well, and then rub shoulders while washing the evening dishes together. The conferences came complete with a wonderful swimming hole. J.J. Lauvergne was instrumental in getting the group started, and Stefan Adalsteinsson, Scott Dolling, and Al Rae were also essential in providing leadership and vision to the group. Several hard-working and clear-thinking colleagues were closely involved in those important early days: Snejana Alexieva, Bernard Denis, Roger Lundie, Xavier Malher, Paul Millar, and Carlo Renieri. I owe a debt of gratitude to all of these people for the organization and insights they provided to this subject. In addition to my COGNOSAG colleagues, Tim Olson deserves a special mention for sharing an interest in cattle color and for freely sharing his insights over several years.

Several individual livestock breeders have collaborated with me over the years. Their contributions have been essential to any progress that has been possible. Colored Angora goat breeders who were essential in helping to crack some of the mysteries of color for that breed include: Sharon Chesnutt, Pat Harder, Isa Jennings, Laurie Lee, and Linda Mercer. Isa always sent large enough samples of the various colored mohair samples so that I could be sure I would be able to hand-spin some into yarn. A mohair sweater in several rare colors resulted from that collaboration, and it still keeps me warm on the coldest winter days. Breeders of other types of goats have also helped: Donna Elkins and Colleen LaMarsh with Pygmy goats, Ed Kinser with Nigerian Dwarf goats, Sandy Van Echo with Oberhaslis, and Sharon Reeves with Myotonic goats. We share tips and leads, which makes the whole effort fun as well as rewarding. Simon Horvat unearthed several fascinating colors in the local Dražnica goat breed, and hiking over the Slovenian Alps with him to see the goats of various colors will always remain a cherished memory.

The color discussion group that formed around Shetland sheep has been especially inspirational and productive. Linda Wendleboe took a few early hunches and transformed them into especially convincing work on some of the lighter colors of that breed, and provided me with a wonderful wool sample which was the result of a multi-year breeding program.

Cattle color genetics varies in relatively few breeds, but interest among breeders of Texas Longhorns (Debbie and Don Davis), Florida Cracker cattle (Stephen Monroe), and Pineywoods (Justin Pitts) has helped move along some of the details for cattle.

Alpaca color has long been of interest to alpaca breeders in North America, and has resulted in collaborations with Denise and Doug Caldwell, Gail Campbell, Bev and Cleve Frederickson, Kenneth Hart, Eric Hoffman, Kenneth Madl, Elizabeth Paul, Nance Sturm, Andy Tillman, and Ingrid Woods. Dan Powell was especially helpful in prodding out some details of llama color. His interest leaked over into the genetics of chicken colors; another fascinating hobby but one that is outside the scope of this book.

Swine color is fascinating because it tends to be so different from that of the other species. Breeders like Cathy Payne and Bryant Rickman have been essential in saving rare breeds such as Guinea hogs and Choctaw hogs. I hope that their experiences will one day solve some of the mysteries of the color variation in those breeds.

Several people have graciously and generously provided photographs that help to illustrate various details throughout this book. Their contributions are noted in the figure captions that accompany the photographs. Those photographs and figures that are unattributed are all my own work.

A major push to actually collect my thoughts and observations into a written work came from Dawie duToit and Trudel Andrag in South Africa. A productive visit there with breeders of Damara sheep, Nguni cattle, and Himba goats made me finally resolve to try to organize the information gleaned from scientific literature, as well as from my own experiences. My hope was that these thoughts might be helpful to breeders interested in the color of their animals. The Damara Sheep Breeders of Namibia were especially instrumental in hosting a portion of my visit. That visit sparked observations that led to my pondering the fine points of color genetics in that highly interesting breed of hair sheep.

Finally, Maria Antonia Revidatti has gently badgered me over the last several years during each of the annual symposia of the Ibero-American group that meets to discuss advances in breed conservation throughout the Americas and Europe.

Putting all of these threads together in a single place has been a long process. My wife,

Torsten Sponenberg, helped that process along through her careful reading and editing. Especially because she is not a geneticist, she greatly helped me to get the information to flow more logically and understandably. Her contributions have made this complex material easier to follow.

The whole staff at 5m Books has been a delight to work with. Their enthusiasm for this project has been contagious. They are always constructive, creative, and thoughtful in their many additions to the final published product, and they are greatly appreciated.

Figure Attributions

The photographs and drawings in this book are by D. P. Sponenberg, with the exceptions noted here. These people have all generously shared their photographs in order to make this work much more complete than otherwise possible. These contributions are gratefully acknowledged.

The Associação de Creadores de Bovinos Mertolengos. Figure 8.40.
Jeannette Beranger. Figures 1.8d, 8.3, 8.10, 8.11, 8.30, 8.36, 8.38, 8.43, 8.46, 8.49, 12.1, 12.2, 12.3, 12.5a, 12.6, 12.7, 12.10, 12.11 and 12.12.
Scott Dolling. Figure 5.19.
Dawie duToit. Figures 4.5, 5.18, 5.23 and 5.30.
Cindy Dvergsten. Figure 5.16.
Leslie Edmundson. Figure 2.28b.
Lyn Hansen. Figure 5.11.
Pat Harder. Figures 2.1c, 4.8, 4.10a, 4.10b, 4.11 and 4.12.
Nancy Irlbeck. Figure 5.4.
Dave Kauffman. Figures 7.5 and 13.1.
Karl Kerns. Figure 12.9.
Ed Kinser. Figures 10.5, 10.12, 10.13, 10.14, 10.18, 11.1 and 11.2.
Elizabeth (Becky) Lundgren. Figure 8.27b.
Roger Lundie. Figure 5.15.
Marie Minnich. Figures 5.1b, 5.6, 5.32, 6.1, 7.9, 7.10, 7.11 and 7.12.
Fabrice Romain Monteiro (© Werner Lampert GmbH) Figure 8.35.
Doug Newcom. Figures 12.8a and 12.8b.
Annabelle Pattison. Figure 2.26d.
Joseph Schallberger. Figure 8.37.
Stone Crest Llamas. Figure 10.11a.
Dr. A. K. Thiruvenkadan. Figures 2.24a and 2.24b.
Vitaliy Tymoshchuk. Figures 5.20a and 5.20b.

Sandy Van Echo. Figure 2.2.
Linda Wendleboe. Figures 5.7b, 5.12, 5.13, 5.14, 7.1 and 7.8.
WoodsEdge Farm. Figures 10.11b and 10.16.

CHAPTER 1

Introduction

1.1 How to Use This Book

The main goal of this book is to give readers an appreciation for the mechanisms of color genetics in six livestock species so they can use the information to assist them in producing colors they desire in their herds and flocks. Knowing how to achieve that goal is important, and understanding the basic organization of the book should help readers to use it most effectively. The material is designed for use by a wide range of readers, all with different reasons for reading it. Many readers will have interests that are limited to only a single one of the six species. Other readers will have broader interests that cover several of the species that are dealt with in the various chapters. Both types of readers will find useful information throughout this book.

The presentation of the information is deliberately organized to lead readers step by step through a subject that is inherently complex. The details of the genetic control of color production can easily prove defeating. A careful approach has been adopted that should help readers through the most complicated issues in order to achieve a good understanding of both basic principles and their consequences as they play out in practical field situations.

This first introductory chapter establishes the broad principles that hold across all six species. It sets out a specific and standardized method for encountering and classifying colors and patterns so that observers are able to use a common nomenclature that allows consistent communication. The following chapters discuss the various species in a sequence that guides readers through the details of the colors of goats, sheep, cattle, llamas and alpacas (together), and then hogs. The order in which the species are discussed is deliberate.

The choice of goats as the first species may surprise many readers. Goats serve well as a general model for most of the others because they have a relatively complete array of variants. Having a good understanding of goat color makes it possible to discuss the other species with greater ease. Among all of these species, the overlap of details is greatest between sheep and goats. Cattle, alpacas, llamas, and hogs each have a slightly different emphasis on which genetic mechanisms are most important for final color production.

Readers with a broad or deep interest will find that reading these chapters in sequence will help them to achieve mastery of many details. However, the material is also designed so that readers interested in only one species can gain the information they need by reading the introductory chapter followed by the chapters addressing their species of choice.

The chapters each discuss a single species and delve into the details of color variation and its genetic control within that species. The first chapter dealing with each species can be thought of as the "what" portion of the text. Each species then has a final chapter titled "putting knowledge to work." Those chapters delve into specific situations or examples that can help readers to grasp the basic information more fully and then use it to achieve their goals. This is the "how" portion of the text. It takes the information and makes it truly useful to breeders. Both goats and sheep have an additional chapter which addresses the rich variation at the *Agouti* locus in these two species. This genetic locus is very important in producing the wide range of easily visible color patterns that goats and sheep express. The *Agouti* locus material is treated separately for goats and sheep because variation at the locus is so extensive that a very detailed discussion is warranted. Readers satisfied with a more superficial coverage may find it beyond their level of interest. The initial more general chapters devoted to goats and sheep do include the basic high points of *Agouti* locus, and this level of attention should be sufficient for many readers.

The unravelling of the mysteries of color genetics in livestock species has included two main routes. One route is the result of careful observation and follows specific variation as it tracks through multiple generations. A second more recent route explores the results of the first route by pursuing and describing molecular changes in the genetic instructions. This has led to an increasing appreciation of the intricacies of genetic mechanisms that contribute to the wide array of colors in all of these species. This knowledge helps breeders to achieve their goals more quickly than was previously possible, and some readers may be interested in the fine details of the molecular basis of the biology of color production. Those details are generally omitted in this book, because they do not contribute very much to the utility of decision-making in most farm settings. The focus of this book is very much on the practical side of being able to use the information for managing colors in herds and flocks on the farm.

1.2 How Color Works in Livestock

Coat color has greatly influenced the breeding of livestock over much of the globe and throughout much of history. The situation today can be misleading, because breed associations of most standardized breeds of livestock have constrained color in those breeds to be minimally variable. This is especially true of production breeds which are the backbone of animal agriculture in developed countries (Figure 1.1).

For example, Angus cattle are black, Herefords are red with white faces, and Holsteins are some combination of black and white. Standardization of color in livestock breeds is so pronounced that most breeders equate any color variation with crossbreeding. The logic behind their assumption is that breeds must have always been pure for color genes, so any variation must come from illicit crossbreeding in the recent past.

Figure 1.1 Most standardized breeds sport only a single color. Brown Swiss cows are brown, European Simmental cattle are red with white faces and body spots. Even in this mixed herd of cattle in Slovenia, those breed-specific colors neatly separate the cows into their breed identities.

The truth is just the opposite. The uniformity of today's standardized breeds descends from a much more variable past, and results from breeders having focused on only a limited array of the more numerous choices they once had. This past variability periodically expresses itself as surprises to purebred breeders and is a reminder of the history behind the breeds of today. The true situation for color variation contrasts sharply to the widely held view of complete consistency for color within pure breeds. Nearly every breed has at least a few hidden genes for color variation. In most breeds the occasional color surprise does crop up, sometimes causing great consternation. These surprise rare offspring that are "off color" can often yield important insights if they are carefully considered and studied.

Some breeds stand out as exceptions to the general rule of standardized breeds having but one color. For example, Nubian goats have always varied in color (Figure 1.2). This may well have been allowed by the registries in the United Kingdom and the USA because the distinctive head and ear shape of Nubians clearly separates them from all other goat breeds available in those countries. Color was not needed to distinguish breed identity because the shape of the goat was unique when compared to other breeds available in the area.

In contrast to the monotonous color choices of many standardized breeds, color tends to be wonderfully variable in adapted landraces and other local breeds (Figure 1.3).

Landrace breeds usually survive in settings that are peripheral to mainstream production agriculture. Landraces are unique by a combination of foundation, isolation, agricultural system, owner attitude, and the type of resources available for sustaining animal maintenance. For landraces, survival has always been more important than uniformity. Indeed, for

Figure 1.2 Nubian goats have a distinctive breed character, so color is not important to breed identity.

Figure 1.3 Damara sheep are a typical landrace, sporting wide color variation within a single gene pool.

several of these breeds, variation was actively celebrated and fostered rather than shunned or avoided. Such breeds are especially helpful in untangling the details of color genetics. Landraces are considered to be interesting not only for their beauty, but also for a depth of cultural relationship formed with their owners that is often lacking in most standardized breeds. Landraces offer a golden opportunity for understanding much about breeds, how they function, and how they can be managed for future usefulness.

1.3 Basics of Genetics

The science of genetics can be unnerving to many people, but is greatly simplified if a few key rules are remembered. Most of the confusion surrounding genetics comes from the words used, and not from the concepts that those words represent. Understanding a few basic concepts greatly simplifies unravelling the details of any genetic system. This discussion does simplify some details, but those details have little to no effect on final color production and so can safely be ignored.

The external appearance of an animal (the phenotype) may or may not reveal all the aspects of its genetic makeup (the genotype). Breeders are usually interested in the genotype of an animal because it affects breeding value due to its contribution to a desired external appearance (phenotype) in offspring. Making a link between the two is the goal of genetics, so that by focusing on the genotype it is possible to assure desirable phenotypes in the offspring.

At the most basic level each individual animal gets half of its genetic material from its sire, and half from its dam. In its own turn, each animal contributes half of its genetic material to each of its offspring. It is always half, but is a randomly different half for each offspring.

The pairing, halving, and renewed pairing of the genetic information at each generational step is possible because the information exists in a duplicated form. Each bit of

Figure 1.4 Chromosomes can be viewed somewhat simplistically as tightly ordered specific sequences of genetic loci that line up like beads on a string. Each locus has its own specific spot on a specific chromosome.

genetic information is one member of a pair, with a few exceptions that are not important to the specific topic at hand. Imagining this juggling of genetic combinations at each generational step makes it possible to greatly increase the odds of producing outcomes that are important to individual breeders.

A few definitions are unavoidable. The genetic information for various traits occurs as discrete pieces. These pieces are called genes and these are made of deoxyribonucleic acid (DNA). The genes occur along chromosomes (Figure 1.4). A useful mental image is to view chromosomes as a sequence of beads (genes) along a string (chromosome). The chromosomes, and their genes, exist in a specific highly ordered and repeatable arrangement across the entire species. The specific location of a gene is called its locus. A locus, therefore, is a sort of address for the physical site of a gene within the genetic material. The plural of locus is loci. The chromosomes occur in pairs, and each member of the pair has a sequence of loci that is identical with its mate.

The specific information in genetic material can vary from individual animal to individual animal. This is the underlying source of variability in the final traits they each express. The alternate forms of genes are called alleles, and each one is caused by a different mutation. The result is that across a species a single locus can host multiple alleles. Importantly, each individual animal is limited to having only two alleles at any one locus, because each animal only has two copies of each gene due to the paired character of the chromosomes. While an entire species might have many allelic variants at a single locus, each individual animal can only have up to two.

Sorting through the effects of the different alleles at the various loci is frequently a complicated task because several loci interact to yield the final appearance of an animal. One useful strategy helps to make order from what can at first appear hopelessly chaotic. First, consider the normal type for wild, undomesticated animals. The mutant alleles can then be considered against this background phenotype, taking each mutant allele one at a time. This approach can simplify understanding, even though the final phenotype of many animals results from the interaction of several mutant alleles at several loci that each change that original wild type. The key to this approach is that each allele can be separately considered and understood for the specific part it plays in the final result of the combination. This approach may seem somewhat artificial when considering animals with obvious and large changes from their wild ancestors, but is really the only approach that can provide insight into the essential fundamental change that each mutation is making as illustrated in Figure 1.5.

The details, as always, become a bit complicated. For most animal species a single wild-type color is designated. This serves well to organize the naming of genes and alleles, and is a useful convention. It does turn out, though, that even in wild species there is usually at least some moderate degree of variation in color. Even though this is the case, it is still

Figure 1.5 Some animals have multiple changes from the original wild type and it can be difficult to tease out the various contributing factors and understand them. This Nguni bull is blacker than wild type, has a white spotting pattern or two, and then speckles of color added back into the white areas.

practical to designate as "wild type" that color (and its genetic machinery) that is most common in the wild species. This convention works well for the species considered here, but keep in mind that for species with a great deal of color variation, such as that seen in wild wolves, the whole issue becomes more complicated very quickly.

The alleles at a specific locus can interact in different ways. One of these is dominance, the flip side of which is recessiveness. Dominant alleles are expressed whether there is one copy or two, while recessive alleles are only expressed if both copies are the same. In a sense, dominant alleles mask recessive alleles when they are paired together. The result is that the recessive allele is unexpressed (hidden) when paired with the dominant one. Importantly, this means that the expression of the recessive allele (or gene – these two words are often used interchangeably) can pop up as a surprise when it becomes paired with another copy by mating animals that both carry the same recessive allele.

When both copies (both alleles) of the gene are identical, the pairing is labelled homozygous. This is true whether the gene is dominant or recessive; the term only refers to the fact that both copies are the same. Homozygous animals can obviously only pass along that one allelic choice to their offspring. In contrast, animals with different forms of the gene making up the pair are labelled heterozygous, and can pass along either one allele or the other to their offspring, as seen in Figure 1.6.

Importantly, the concept of dominance and recessiveness only describes the interaction of alleles at a single locus. It does not relate at all to the frequency of the alleles in a breed. "Frequency" specifically refers to how often each allele occurs in the population being considered. Some breeds are indeed uniform for recessive genes, such as the Red Poll cattle breed and its red color. In contrast, some rare alleles are passed along as dominants, such as the gene that causes nearly white Jersey cattle. It is important to not confuse allele frequency with dominance. The status of an allele as dominant or recessive does not change over time, even though its frequency can. Even if a recessive allele is uniform throughout a breed, it is still recessive. This becomes obvious in crossbred offspring where the expression of a recessive allele becomes hidden when the other parent breed has a dominant allele at the same locus.

Pairing up standardized breeds with an appreciation for their uniformity of dominant and recessive alleles is actually a common strategy for clearly identifying crossbreds

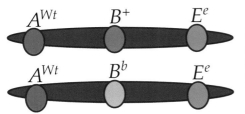

Figure 1.6 Chromosome pairs can vary in their specific variants. In this example, the *A* locus is homozygous for the dominant allele A^{Wt}, the *B* locus is heterozygous for the *wild type* (B^+) and *brown* (B^b) alleles, and the *E* locus is homozygous for the recessive E^e allele.

between two breeds. For example, horned Hereford cattle are homozygous for recessive red, recessive presence of horns, and dominant white face. Angus cattle are homozygous for dominant polled (lacking horns), dominant black, and recessive "no white face." Mate the two together, and the three dominant characteristics (polled, black, white face) are consistently expressed in the resulting hybrid offspring, while the three recessive genes (horns, red, no white face) seem to disappear. Those three recessive alleles are present, but are hidden from external expression by virtue of being paired with dominant alleles. Mating this crossbred generation together produces the original combinations along with several others as the alleles assort themselves out into all possible combinations (Figure 1.7).

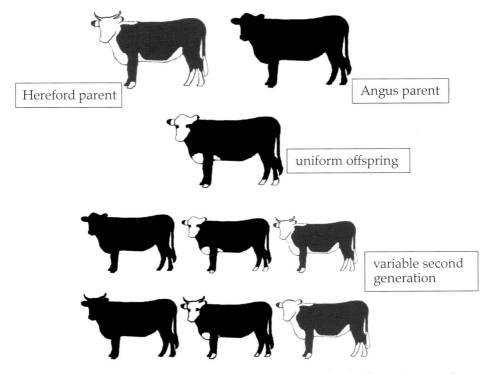

Figure 1.7 Horned Hereford cattle are uniform for horns, red, and white faces. Angus cattle are uniform for polled, black, and no white faces. Crossing these two breeds yields a uniform crop of offspring that express the dominant alleles of both parent breeds. If those offspring are used among themselves, all combinations of the three traits appear, including some that were not present in the original parents.

Two important concepts involve situations that are intermediate between the usual dominant-or-recessive pattern. These are codominance and incomplete dominance, and they function slightly differently from one another. Codominance occurs where both alleles are expressed when paired together. In a sense, neither is dominating the other, but both are expressed. Incomplete dominance is slightly different, and occurs where two, one, or no copies of an allele each have a recognizable result in the appearance of the animal. In most cases of incomplete dominance each homozygote has a specific appearance, and the heterozygote usually has an appearance intermediate between the two different homozygotes. This can be thought of as three levels of expression of the same basic trait, in contrast to codominance where both alleles are expressed reasonably fully.

Codominance is present in a few coat color loci, notably the *Agouti* locus, but is otherwise fairly rare in genetic systems outside of blood types and other protein systems. Incomplete dominance is more frequent, and occurs in several loci controlling coat color. Incomplete dominance is especially common at loci controlling the dilution of color. In those systems, one homozygote usually has a fully intense or dark color, the other homozygote is very pale, and the heterozygote is intermediate between the two extreme shades of color. Similar relationships are found for some of the alleles that cause white spotting, where one homozygote lacks any white, the other is nearly entirely white, and the heterozygote has an intermediate level of expression of white areas that is easily recognizable. The underlying genetic mechanisms are slightly different between codominance and incomplete dominance, but these two systems both have similar advantages to breeders because heterozygotes and each of the homozygotes can be detected accurately from visual inspection alone (Figure 1.8).

The interactions of alleles at a single locus are usually relatively easy to understand. Alleles and genotypes (specific combinations of alleles) also interact across multiple loci which is more complicated. In some situations certain genotypes at one locus can mask the expression of information at a second locus. This is somewhat like dominance and recessiveness, but occurs across multiple loci instead of at only a single locus. In this situation the genotype causing the masking is referred to as epistatic, while the one that is masked is hypostatic.

In a general sense, both dominance and epistasis can be viewed as mechanisms by which a part of the genome is masked from view. The masked information can be revealed by targeted breeding strategies, although it can also pop up as surprises in a number of breeds following the mating of two animals that have such masked information. Genetics is a fascinating way to attempt to unravel the hidden potential of animals to produce desirable phenotypes – or to avoid producing undesirable phenotypes. Understanding that dominance and epistasis can mask portions of the genetic material is a huge step in understanding how to use genetics to achieve desired results by either unmasking genetic information, or assuring that it remains masked and unexpressed.

Many people hope that genetics will be the key to assuring specific outcomes for each and every mating decision. While this can be true in some situations that involve minimal genetic variation, it is more usual that genetics simply serve to constrain the possible outcomes to a relatively narrow range. This can still be useful for breeders, but focusing on the idea that the final outcome is within a narrow range is very different from

Figure 1.8 Codominance and incomplete dominance are fundamentally different, but have similar advantages to breeders. The codominant tan regions of the Myotonic goat patterns in A and C are each fully expressed in B (ignore the white spots!). The Pineywoods cow in D lacks an incompletely dominant dilution allele and is fully intense red (photo D by J. Beranger). The calf in E is a fully dilute homozygote, while its dam's spots of an intermediate yellow color indicate that she is heterozygous.

viewing genetics as a tool to assure a very specific outcome for each and every animal that is produced.

1.3.1 Genetic Nomenclature

Genetic nomenclature can be a real headache. The Committee on Genetic Nomenclature of Sheep and Goats (COGNOSAG) has made great strides in standardizing nomenclature for different species of farm animals. Several conventions help in the effort to make nomenclature uniform and understandable. One is that locus and allele names are always in italics, while phenotypes are not in italics. For example, *dominant black* is an allele, black is a phenotype. Locus and allele names usually try to convey the general character that is affected. Locus names begin with a capital letter (*Agouti*), while allele names are always lower case throughout (*black and tan*).

In order for discussion to proceed more easily, abbreviated symbols for loci and alleles are often used. These are in italics, and the locus abbreviation is capitalized. The allele abbreviation follows the locus as a superscript, and it is usually not capitalized unless it corresponds to a dominant allele. This can be confusing, because when spelled out the dominant alleles always have a lower-case initial letter, even though the symbol may have an initial capital letter. By this method A^{Wt} is the *white/tan* allele at the *Agouti* locus, and the capital letter of the superscript for the allele indicates that this is the dominant allele at that locus. The *wild* allele at each locus is designated as + which avoids confusion with other alleles. This is the allele from which all the others descend as mutations.

A modern trend is to increasingly resort to mouse genetic nomenclature when referring to most loci. This might be disorienting at first, especially to those with some background in previous genetic nomenclature for farm animals. The rationale for resorting to the mouse is that its genome is the best understood of all animals, and by using the mouse names for loci the homologies of action across species can be better appreciated. Where possible, in this book the older locus names will be correlated with the mouse or biochemical names. This book generally retains the older, more traditional names for loci because many readers are already familiar with them.

1.4 Basics of Pigmentation

Animals are colored due to pigments in hair and skin. Mammalian skin and hair have two pigments, eumelanin and pheomelanin. The presence and character of these two is controlled differently by different genes. It is therefore essential to identify which pigment is present in a specific region of an animal in order to understand the genetic mechanisms that are active in yielding the final color.

Eumelanin is made from the amino acid tyrosine. Unmodified eumelanin is black, but it can be modified to dark brown, flat brown, slate-blue, smoky, or beige. As a very general rule all of the eumelanin on any one animal is a single final color, so that animals can have black eumelanin, brown eumelanin, or slate blue eumelanin but not two or three of the choices. This generality tends to fall apart a little in fiber-producing animals because of the manner in which long fibers take up melanins. Still, as a general rule it is true to say that individual animals have just one form of eumelanin.

Pheomelanin is made up of both tyrosine and cysteine, and is usually referred to as "tan" but frequently runs a range from very dark brown to chestnut red, tan, gold, and all the way to cream. It is fairly common for pheomelanin on a single animal to have different shades and tones across different body regions. In this regard it is quite different from eumelanin in the way it reacts to genetic control across the body.

The biochemical differences in the two pigments are complex, and the main importance of this detail is to indicate how fundamentally different these two pigments are. Understanding that they are different helps to explain how differently they interact with

so that the entire fleece converges on a narrow range of fairly coarse diameters. The result is a fleece of uniform diameter and lustrous sheen. Other fleeces, such as the coarse-wool and carpet-wool types, tend to retain differences between the diameters of primary and secondary fibers with both types growing fairly continuously but at rates that are different for the two types. This yields long coarse and short fine fibers. In all types of fleeces the pigment tends to become distributed less densely than it is in unmodified short hair coats. The result is a fleece that is somewhat paler than a short hair coat. Exceptions to this loose rule are relatively common, though, and some fleeces are fully and intensely pigmented. This is especially true of eumelanic colors, and is also especially true of llamas and alpacas.

In contrast to most sheep fleece types, the fleeces of goats, llamas, and alpacas can retain pigment very well. The result is fleeces that are in many cases more darkly pigmented than are the usual variants in sheep fleeces. Suri alpacas have yet another distinct change. Suri fibers have large flat superficial scales, which leads to the extraordinary slickness and sheen of the fiber.

1.7 Basic Color Identification

Considerations of genetics aside, a first step in evaluating color is to have accurate identification of color. This sounds very much easier than it actually is. For communication on color to proceed, everyone involved needs to be looking at animals in the same general way so that they are assured of using the same general vocabulary. This approach avoids misunderstanding at the level of color identification, which is essential if underlying genetic mechanisms are to make sense. Unfortunately, there is no standardized English vocabulary for the colors of all of these various species of animals, and the situation becomes even more confusing when other languages are considered.

Accurate identification of color and consistent use across different observers is an essential key for successful studies that correlate specific phenotypes with underlying genetic mechanisms. Lack of a consistent color nomenclature has defeated effective genetic studies in many species for which they would be very useful to breeders. A standardized nomenclature is therefore a first step in understanding color classification.

One approach that works well is to first decide on the basic color of the animal in question. Basic colors can usually be broken down into black, tan, some combination of these, or some modification of these. This is, of course, after eliminating entirely white animals that defeat this step.

The combinations of black areas and tan areas usually occur as distinct repeatable patterns that vary species to species. Each pattern has its own unique name, often reflecting the peculiarities of the pattern, its breed of origin, or its geographic location. This is especially true of sheep and goats, and somewhat less so for cattle, llamas, and alpacas. Each species varies in the extent to which the patterns each have a specific name, but striving for uniformity in naming these greatly helps the understanding of the mechanisms and genetics of color. The names suggested in this book are not uniform across all breeds or across all species, but can still serve as a reference point for the discussion of their genetic control. The patterns of black and tan areas are especially important in goats and sheep. Unfortunately,

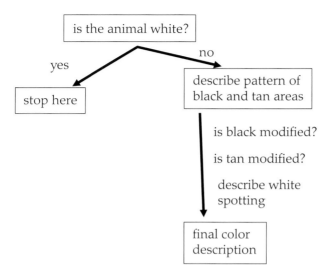

Figure 1.15 Using a standard flow of decisions on color can help to achieve a consistent identification of color in all of its details.

the literature has tended to transpose these from one species to another, despite the fact that in most cases the patterns in one species are only similar rather than truly identical to those in another species.

After the basic pattern of black and tan areas is documented, then any modification of these two pigments can be considered. This step completes the definition of the basic color. After deciding on the basic color of the animal, any patterns of white can be noted. This will be more difficult to do on animals with combinations of several patterns, but any attempt is usually reasonably successful and is certainly better than ignoring details or lumping all "white spotted" or "roan" animals into one large group. The overall flow is outlined in Figure 1.15.

1.8 General Principles Across All Species

The fine details of color genetics for each of the six species (goats, sheep, cattle, llamas, alpacas, and hogs) are distinct enough that each requires a separate discussion. However, a few overarching principles hold true for all of them, and are useful as a beginning point. The interactions of the *Extension* locus with the *Agouti* locus are especially important.

1.8.1 Extension *Locus*

The *Extension* locus (*Melanocortin 1 Receptor* or *MC1R* in mouse nomenclature) is the first major control site for pigment production. This locus determines the types of melanin an animal can produce, and its name derives from "extension of black" which reflects its basic action in several species. In most species the *Extension* locus can act in three different ways. Distinct alleles are the means through which this action occurs. These three possible actions at *Extension* are important because each constrains the range of actions that are possible at other loci. The color variation in most species relies first on the specific alleles that are present at *Extension*.

The most dominant allele is *dominant black* (E^D). This allele is well documented in sheep and cattle, and less so in goats and alpacas. The *dominant black* allele results in a completely black phenotype unless modified by other loci. If the only change from wild-type color is *dominant black*, the result is a completely black coat, although some animals will be "imperfectly black" with either rusty black or very, very dark brown as their final color. In general, though, all such animals would be classed as black by most observers.

The *dominant black* allele is epistatic to the other most important color-determining locus, *Agouti*, which will be discussed in the following section. This *dominant black* allele therefore achieves a solid black (eumelanic) phenotype even in the presence of other information at the *Agouti* locus. The *dominant black* allele is a common source of black cattle, sheep, and hogs. It is rare in the other species, but does occur.

A middle allele at *Extension*, in terms of dominance, is *wild* (E^+). This allele, as the name suggests, is present in wild animals of all six species. This allele allows the *Agouti* locus to be expressed, because it is not epistatic to that locus. Most *Agouti* locus patterns have both black and tan areas, although a few exceptions are important and will be discussed both as generalities with that locus and with each species individually. It is the presence of both black and tan areas that tends to betray this E^+ allele in most genotypes of most species. All of the species have this allele, although it is more rarely used as a source for colors of standardized breeds of cattle and hogs than it is for the other species.

The recessive allele at *Extension* is *red* (E^e). This allele results in a uniform (or nearly uniform) tan or red phenotype with no black areas. In cattle this is the familiar "red with a white switch" of the Red Angus or Red Poll breeds, and also the Hereford (with a specific pattern of white spotting added). This allele, though recessive at this locus, is also epistatic to *Agouti* so that when an animal is homozygous E^eE^e the result is red regardless of what *Agouti* locus instructions are present. The important detail here is that the animals are entirely red instead of combinations of red and black.

The *red* allele is common in cattle as a source of red cattle, but occurs so rarely in sheep that it has limited distribution. It is also rare in goats, occurring in both Oberhasli and Angora goats, as well as in breeds not currently in the USA. A few llamas and alpacas also have this allele. While it may be present in these species other than cattle, it is certainly much rarer in them than it is in cattle. In contrast, it is widespread in cattle and is frequently used as a source of breed-specific solid red color in several breeds.

The interaction of these three alleles organizes much of the basis from which other colors are built in most species (Figure 1.16).

1.8.2 Agouti *locus*

The *Agouti* locus (*Agouti Signalling Protein* or *ASIP* in mouse terminology) governs the distribution of black and tan areas. The name comes from a South American rodent with a distinctive color pattern. In most of the species considered in this book the patterns are symmetrical. Indeed, many geneticists reserve the word "pattern" for the *Agouti* locus patterns, although this causes a problem with what to call the different results of white spotting alleles if they are not "patterns!" General phenomena of the *Agouti* locus are discussed here, with more detailed investigations of specific patterns discussed separately for each species. While generalities do hold true across the various species, each species has its own

Figure 1.16 The *Extension* locus in cattle yields the three basic colors from which others are built: black (A) is dominant, wild type (B) is a combination of tan and black areas and is intermediate, and red (C) is recessive. All photos are of Criollo Patagónico cattle.

peculiarities in some specific *Agouti* patterns. A full understanding must rest on a separate consideration of the details for each species.

The *Agouti* locus has a few consistent characteristics across species. One characteristic is that several different alleles occur in each species. Goats and sheep each have well over 25 different alleles, and each one of them causes a distinct symmetrical arrangement of tan and black areas. A second important detail is that all tan areas are consistently expressed in heterozygotes. The practical result of this is that the more extensively tan patterns are dominant to the more extensively black patterns. This trend yields a somewhat loose array of dominance relationships among the alleles. Completely tan is at the top of the dominance hierarchy, even though the pattern caused by this allele has very minor black regions in some species. Next in line of dominance are the patterns that have symmetrical arrangements combining both tan and black regions. The most recessive allele is completely black caused by the *nonagouti* allele (also called *no pattern* in some references). When animals have two of the intermediate patterns, the tan areas of each pattern are expressed. Remember that each animal can have up to two of these alleles, but never more than two. The combining of patterns can result in some interesting colors that can be confusing unless the observer is familiar with the consistent expression of tan areas as determined by *Agouti* locus alleles. The intermediate alleles are accurately thought of as codominant to one another, where the tan areas of each are fully expressed in the heterozygote.

Heterozygotes that carry *nonagouti* tend to have somewhat darker but more distinct

Figure 2.1 The three different coat types of goats influence the final appearance of many colors. The shorthaired Myotonic goats in A have the advantage of clearly showing the distribution of tan and black areas. Cashmere goats, such as the Myotonic doe in B, typically have pale fine fiber that contrasts with the darker background color. Most Angora goats, even black ones, grow mohair that is a pale grey like the recently shorn goat in C (photo C by P. Harder).

foundation of all these other sources of variation. This discussion first focuses on color in shorthaired goats, which can be best understood by taking the various genetic factors into consideration in a sequence that cascades down through their influence on the final color. Practical aspects of managing color in fleece-bearing goats are more fully discussed in Chapter 4.

The general control of color in goats is through only a handful of different genetic loci. The *Extension* and *Agouti* loci set the stage for a basic color phenotype, and then various other modifying factors provide for a final color. Table 2.1 introduces the factors contributing to the basic goat colors.

2.1 Extension *Locus*

A *dominant black* (E^D) allele does occur in goats, although it is not yet documented as occurring at the *Extension* locus. For convenience it can be considered to reside here because the practical implications to its actual location are minor. A *dominant black* allele has been noticed in colored Angora goats, and breeding results indicate that it also occurs in other Asian, African, and perhaps a few European breeds. Goat color has been poorly studied in most regions, so it would not be surprising to find *dominant black* in a number of breeds as a rarity. The result of the *dominant black* allele is a completely black coat, and this masks any *Agouti* locus expression.

Table 2.1 Factors contributing to basic goat color.

Extension locus	*Agouti* locus	dilution factors	other
dominant black (rare)		*brown*	moon spots
wild	pattern groups:	tan dilution	
	nearly tan	others (rare)	
	tan with black trim		
	tan with black periphery		
	black front, tan rear		
	tan front, black rear		
	black with tan trim		
	nearly black		
red (rare)			

Most goats have *wild* (E^+) as their *Extension* allele, and this allele allows the expression of the *Agouti* locus. The practical result of this is that the *Agouti* locus is the main determiner of color in most goat populations. Most *Agouti* alleles cause patterns that have a bilaterally symmetrical interplay of tan and black areas. The symmetry of the patterns provides for easy assignment to the *Agouti* locus, which in turn also betrays that E^+ is involved.

A recessive *red* (E^e) allele occurs rarely in some breeds. This allele is likely to reside at the *Extension* locus. It occurs in Oberhasli goats, Angora goats that are selected for color, and some Nigerian Dwarf goats (Figure 2.2). Recessive red from the E^e mechanism can phenotypically resemble the much more common red produced by the A^{Wt} allele at the *Agouti* locus. However, the two types of red behave very differently in breeding programs. The recessive E^e red can be produced by two parents that each have obvious *Agouti* patterns, which is impossible for the red goats caused by the dominant A^{Wt} allele.

Figure 2.2 Red Oberhasli goats are most likely due to a recessive *Extension* locus allele (photo by S. Van Echo).

This reflects the fact that E^e is recessive but is also epistatic to the *Agouti* locus. Red goats with A^{Wt} must have a red parent, which is not necessary for the E^e reds. This *red* allele at *Extension* would be particularly useful in achieving a uniformly red population, because a population homozygous for a recessive allele can produce no other colors as surprises. The red of these goats is generally very uniform, with little of the light or dark trim or shading across the body that occurs on many red-based goats resulting from various *Agouti* locus alleles.

A key point is that variation in the alleles at the *Extension* locus is very rare in goats, and generally occurs in only a few breeds. In most situations this locus can be safely dismissed as contributing to color determination, with the exception of those few breeds. However, in those few breeds or herds where variation at *Extension* does occur, it becomes very important in determining the genetic control of color. In those situations the *Extension* locus is the main first choice in the final color, and therefore cannot be ignored.

2.2 Agouti *Locus*

The *Agouti* locus is the source of most variation in goat color in different breeds worldwide. Most color variation can therefore be easily understood by appreciating the intricacies of how the *Agouti* locus functions to produce colors. The fine details of the *Agouti* locus can easily become overwhelming because there are over 30 distinct alleles, many of which lead to only subtle differences in the final symmetrical pattern of tan and black areas. These numerous patterns neatly sort themselves into a few major groups, and this helps in understanding the locus at a fairly basic and broad level that serves most breeders well. Considering the patterns as residing in a few major groups is the approach taken in this chapter. The finer details are explored more exhaustively in Chapter 3 for readers that are eager for a more detailed exploration of the locus and its effect on color.

Several breed registries have used one or a few of these distinct patterns to define breed identity. A handful of other standardized breeds, such as the Nubian, have allowed wide variation to persist. Many of the landraces that are not tightly standardized allow a great variety of *Agouti* locus patterns and therefore many landraces sport several different coat colors.

The alleles producing the *Agouti* locus patterns have an interesting distribution across goat breeds. While many goat breeds have only a single officially recognized *Agouti* locus pattern, the converse is rare: it is quite unusual for a specific *Agouti* pattern to be limited to only a single breed. Many observers incorrectly infer that the specific patterns that have been used for breed standardization occur only within a certain specific breed. For example, the combination typical of Toggenburg goats occurs in several breeds, and such goats are often confused with Toggenburgs despite a lack of any influence from that breed. Separating color from breed origin is a difficult concept for many goat breeders, but it does reflect the reality of the colors and their origins. Basically, the genes for color existed well before standardized breeds did, and not the other way around. Finding one of the more breed-specific patterns in a goat population is not evidence, by itself, that the population under consideration has necessarily received genetic influences from the standardized breed with the pattern.

Another peculiarity of the breed distributions of the *Agouti* locus alleles is that many of them do indeed have a very limited range. Many alleles are limited to only a few breeds, raising the possibility that these breeds are indeed related to one another. In contrast, several of the alleles occur in breeds of very divergent origins and occur worldwide. It is tempting to propose that these more globally encountered alleles hail back to mutations that occurred in the distant past and have had more time to be shared widely among goat populations.

Most *Agouti* alleles produce strikingly symmetrical patterns of tan and black areas that are the same from side to side on the goat. Most of these patterns are readily recognizable and can be clearly distinguished from one another. Unfortunately, it is also true that a few of the patterns caused by single, defined *Agouti* alleles vary over a relatively large range rather than being consistent over a narrow and easily recognizable range. Recognizing and identifying these more variable patterns is challenging. Fortunately, the fine details of face, ear, and leg markings in the various patterns are generally highly repeatable among all goats with a specific pattern. This trend holds true even for those patterns that are highly variable. Emphasizing those specific areas while evaluating colors can help to untangle what can be potentially confusing patterns.

Most of the variation in *Agouti* patterns can be understood as changes to two different aspects of any single pattern: the relative shade of the tan areas, and the relative extent of the black areas. Tan areas vary from dark red to light cream. Black areas vary from extensive to minimal. Some specific individual *Agouti* alleles produce quite different final patterns with the palest tan and least extensively black goat at one extreme, and the darkest tan and most extensively black goat at the other. Several alleles can produce opposite extremes that are barely recognizable as being the same *Agouti* locus pattern. Multigeneration observations in families of goats with these alleles have demonstrated that these ranges of pattern are indeed due to single *Agouti* alleles (Figure 2.3). Fortunately, most of the patterns produced by alleles at the *Agouti* locus behave more predictably and are easily recognizable.

The color of tan areas on goats varies in a few different ways. The most usual variation is from creamy white, through tan, to a dark tan. A second range of tan colors involves a more distinctly red character, and the progression in this situation is from cream, through gold, to red. Sometimes the very dark shades of tan have a brown character rather than the usual red character that is more typical of pheomelanin. These darker shades of tan can often be confused with brown eumelanin. Variation in pheomelanin expression can dramatically change the appearance of the *Agouti* patterns on which they occur.

The shade of tan is under genetic control, and in most populations the control seems to be fairly simple and involves only a few genes. The alleles that modify the shade of tan are not at the *Agouti* locus but do modify the tan areas of the patterns at *Agouti*. Many breeds of goats have experienced past selection to make the goats a reasonably uniform color pattern, with the result that most breeds do not have wide variation in depth of tan. However, some breeds do retain variability and the patterns can be confusing unless the observer remembers that the depth of tan can vary. Because the shade of tan can be so variable, the overall distribution of black areas is often the most reliable key to the accurate recognition and classification of most *Agouti* patterns. Unfortunately, even the distribution of black areas can be misleading in some patterns.

Figure 2.3 These *peacock* pattern Myotonic goats all have the same *Agouti* locus genotype ($A^{pck}A^a$). The variation in final color is due to differences in shade of tan and extent of black. Goat A has the palest tan, and the least black. Goat B has pale tan and extensive black. Goat C has dark tan and extensive black. Goat D has dark tan and less extensive black.

It is also essential to note that while the shade of tan does vary for several *Agouti* locus alleles, many of the *Agouti* locus alleles have a somewhat restricted range of shades of tan. Basically, some of the alleles tend to produce tan regions that are distinctly dark shades, while other alleles produce tan areas that are generally at the paler end of the range. Some alleles seem restricted to the pale–gold–red shades, while others are restricted to the pale–light tan–dark tan shades. To add to the complexity, some *Agouti* locus alleles seem capable of hosting the entire range of tan shades. Goats that are heterozygous for two *Agouti* alleles with different potentials for the shade of tan can have dramatically patterned coats. The final result depends on which specific tan areas are unique to which of the patterns in the combination because they usually only retain the depth of tan for that specific pattern rather than the other one (Figure 2.4). Despite this tendency for each allele to have a restricted range of shades of tan, some individual goats flaunt this and can express a shade that is unusual for the specific *Agouti* allele. This happens frequently enough that a general and unbreakable rule is impossible to establish.

The second dimension for variation in *Agouti* locus patterns is the extent of the black

Figure 2.4 The Myotonic kid (A) is heterozygous for *peacock* and *blackbelly*. The combination has resulted in cream regions where *peacock* is expressed, darker tan regions where only *blackbelly* is expressed on the rear of the body, and black where both patterns are black. The Myotonic buck (B) is heterozygous for *toggenburg* and *tan cheek*. The pale tan areas of *toggenburg* contrast with the dark tan of *tan cheek*, and both are fully expressed.

regions. Fortunately, in most patterns the extent of black regions does not vary widely, resulting in patterns that are easily recognizable because they are so repeatably uniform from goat to goat. A few, though, are hypervariable and produce individual goats that vary from nearly all black to nearly all tan. In the middle of the range these patterns are usually very distinctive and easily recognizable, but at the extremes they are easily confused with patterns caused by other alleles. Knowing how these patterns vary from generation to generation can be the only clue that a single allele is at work, and this poses real challenges if color is assessed and classified at only one time and with a single generation.

The goat *Agouti* patterns can be fairly neatly sorted into broad groups based on the predominant pigment that is present as well as the distribution of that pigment over the body, as presented in Table 2.2. The fine details of the patterns can take the identification of the patterns to a very exacting level of description. That level of detail is presented in Chapter 3. The final list of *Agouti* alleles is quite long, and new patterns continue to be described as they come to light. The extent to which this ever-expanding list results in practical differences in color is debatable, even though chasing down this source of variation in color has turned out to be a fascinating endeavor for many people. Several of the patterns are very similar to one another and are therefore of more theoretical than practical interest. The overarching key is their predictability and the potential for breeders to use them to create a desired final color.

At one extreme of the pigment distribution the *Agouti* locus patterns consist of those that are predominantly tan. Most of these patterns have moderately to highly variable expression in both the extent of any black areas and the final shade of tan areas. The consequence is that exactly where to draw the boundaries between these patterns is problematic. The "least extensively black" extreme of all of them converges on uniformly white, tan, or red, with the final shade depending on what is happening in the "depth of tan" dimension of

Table 2.2 Major groups of *Agouti* locus patterns in goats, along with subgroups of some of them.

nearly tan	tan with black trim	tan with black periphery	black rear, tan front	black front, tan rear	black with tan stripes	black or nearly black
		pale group dark group			belly black belly tan	

variation. The "most extensively black" end of the spectrum is usually more distinctive and more easily identifiable because these black areas are different from pattern to pattern.

The most widespread allele in this group is *white/tan* (A^W) which is the most dominant *Agouti* allele (Figure 2.5). It causes a reasonably uniform tan coat, which can be stark white, gold, tan, or deep red. A few other alleles in this group of mostly tan patterns consistently add relatively minor black or cream areas, but all of them can be generally classified as tan.

Figure 2.5 The *white/tan* allele in Myotonic goats can vary in expression from starkly white (A) down to red or mahogany (B). Some goats with this allele have shading over the body, either light (C) or darker (D).

Figure 2.6 The *bezoar* allele causes the pattern typical of wild goats and occurs in several breeds including the Myotonic goat. This pattern has shaded tan areas, along with black and cream trim.

The next group of patterns includes those that are predominantly tan and that have obvious black trim. The *bezoar* (A^+) allele (Figure 2.6) is the main one that characterizes this group. This is the wild-type pattern for goats. It has a shaded tan body, pale cream belly, and black markings on the face, dorsal stripe, and lower legs. Goats with the *bezoar* allele are born a bland tan and acquire the black pattern after a few months. Males are dramatically darker than females and darken with age. A few other alleles create patterns that closely mimic the *bezoar* pattern but lack the typical shading of the tan areas.

The group of patterns with a tan body and black periphery includes the *blackbelly* (A^{bb}) pattern, which is widespread geographically and occurs in many different breeds (Figure 2.7). This pattern is basically a black periphery framing a tan body. The body usually has a fairly uniform shade of medium to dark tan, and some goats are distinctly red. Black covers the lower legs, belly, perineum and bottom of tail, forms a stripe down the back, down the withers and shoulders, and up the bottomline of the neck, and forms an "H" pattern on the head. This group of patterns has one subgroup with several patterns that are relatively pale and have subtle differences in the black regions. A second subgroup includes several patterns that are darker than *blackbelly*, with more extensive black areas, and with dark tan and black mixed into the tan body regions.

Several distinctive patterns have a black rear and a tan front. These are superficially similar to one another, and all have tan on the anterior portion of body and neck and tan on the abdomen. The *peacock* (A^{pck}) pattern is the most common of these. It has a black rear half, black lower legs, and a black belly. The black region on the rear half of the body varies

Figure 2.7 The *blackbelly* pattern of this Myotonic doe is widespread throughout the world.

greatly in the extent of coverage (Figure 2.8). This pattern can vary considerably in the shade of tan and the extent of black. A curved black bar on the lower cheek of this pattern is missing in the other patterns of this group.

Several patterns are nearly the reverse of the previous group, and have a black front and a tan rear. One of the patterns in this group occurs worldwide. This is the *san clemente* (A^{scl}) pattern illustrated in Figure 2.9. The final phenotype of this pattern varies widely, from a "black and tan" type to a "black face" type and everything in between. Other patterns in the group have subtle differences in the final distribution of the black and tan areas.

A large group of patterns are predominantly black and have tan striping on the face, legs, or both. As the *Agouti* patterns become progressively more extensively black, the tendency to vary diminishes. Within this group a few patterns have a black belly. The *toggenburg* pattern is the most widespread of these and occurs worldwide. The *toggenburg* (A^{tgg}) pattern has a black body and belly, with distinctive pale tan (usually cream or nearly white) lower legs, eye stripes connecting to a pale muzzle, and pale upper ears. Within the subgroup of those with a tan belly is the *black and tan* pattern, with a similar worldwide distribution in several breeds (Figure 2.10). The *black and tan* (A^t) pattern has a tan belly, small patches above the eyes and perineum, and legs with black markings that break at the knee.

Figure 2.8 The black posterior and tan anterior is typical of several patterns including *peacock*. The *peacock* pattern has a distinctive curved black bar on the cheek as evident in this Myotonic doe.

Figure 2.9 The Myotonic doe in this photo shows a *san clemente* pattern, which is common worldwide, and is very nearly the reverse of the *peacock* pattern.

The other patterns have more restricted breed distribution. The many patterns in this group tend to be fairly similar to one another, so much so that several standardized breeds include a few of these different patterns without any concern that they could lead to confusion as to breed identity because they are all generally black with tan trim.

The last group includes patterns that are all black or nearly so. The patterns in this group that do have tan areas lack the distinctive leg patterning and facial stripes that characterize the previous group. The most recessive of these is *nonagouti* (A^a), with no tan in the coat (Figure 2.11).

Goats that have two different *Agouti* locus patterns express the tan areas of both patterns. For example, a goat with *peacock*, along with *black and tan*, has the *peacock* pattern, but with the tan eyebrows, belly, and striped legs of the *black and tan* pattern added to it (Figure 2.12). Some heterozygotes can be quite difficult to identify unless the parents are

Figure 2.10 The Myotonic buck in A has the *toggenburg* pattern with a black body with cream trim. The black belly and tan lower legs are typical. In contrast, the Myotonic doe in B has the *black and tan* pattern, with a tan belly, distinctively marked lower legs, and a more modest facial pattern.

Figure 2.11 This Myotonic kid shows a fairly uniform black color resulting from the *nonagouti* allele. This color is very widespread.

known. For example, *san clemente* added to *peacock* can result in a goat that appears nearly all tan. Another example includes *blackbelly/black and tan* heterozygotes that closely resemble bezoar goats. Heterozygotes for *peacock* and *blackbelly* can have cream fronts (peacock) and dark tan rears (*blackbelly*) with black legs and facial stripes retained where neither of the patterns in the combination has tan areas. The key here is that areas that would be black in the *peacock* pattern but tan in the *blackbelly* pattern are deep tan, while those areas that are cream in the *peacock* pattern remain cream. The cream modification of tan areas overrides the areas that would be darker in other patterns.

Figure 2.12 This Myotonic buck has both *peacock* and *black and tan* along with white spotting. The tan areas of both patterns are evident, especially on the legs, belly, and head where the *black and tan* gets added onto the *peacock* pattern.

Another phenomenon of the *Agouti* locus is that goats that are homozygous for the patterns tend to be more extensively tan than those that are heterozygous for *nonagouti*. This trend is not consistent enough to be uniformly true in every individual goat but is certainly true across reasonably large numbers of goats.

2.3 Dilution

Various alleles, other than those at *Extension* and *Agouti*, influence the final color of goats. Most of these other influences make the basic color paler. They can be considered together as forms of dilution. Goats have a few different sources of dilution, but generally have fewer sources of dilution than the other species do.

The *Brown* locus is one common cause of lighter colors. The *Brown* locus alleles act by changing the way that eumelanin is made and packaged into melanosomes. The practical result of this is that all eumelanin over the entire goat is changed, and all pheomelanin is unchanged. This locus has several alleles, including *dark brown* B^{Db} (most dominant), *light brown* B^{lb} (next most dominant), *wild* B^+ (next most dominant), and *brown* B^b (recessive to all of the others). This order of dominance can easily be confusing. The black areas in goats with the *wild* allele remain black, while the other alleles each change all of the black on the goat to produce some shade of brown even though some of these alleles are dominant and one is recessive. These alleles can occur with any of the *Agouti* locus patterns, changing the expected black regions to be brown instead.

The *dark brown* allele, as the name suggests, results in a dark chocolate-brown color (Figure 2.13). This can be subtle on some *Agouti* locus patterns, especially if pheomelanin is dark and close in shade to the brown color caused by this allele. Most *dark brown* goats are born with the eumelanic areas very slightly "off black," and these kids are easily confused with black. As the kids grow, after a few weeks the color difference becomes more obvious. On adults the dark brown is usually very obvious and difficult to confuse with black. Homozygotes are somewhat lighter than heterozygotes but are still dark.

Figure 2.13 This Myotonic kid is *dark brown* over a *black and tan* base from the *Agouti* locus. The effect of the *dark brown* allele is easily seen as affecting only the black regions, not the tan regions.

Figure 2.14 These Myotonic does are light brown in color. The final phenotype of *light brown* can overlap with *dark brown* especially the *dark brown* homozygotes.

The *light brown* allele is the likely cause of the final color of Toggenburg goats, which have the *toggenburg* allele at the *Agouti* locus. The black areas are modified by this allele and change to a lighter and obvious brown (Figure 2.14). The pheomelanic areas remain cream colored, and the combination of colors produces the hallmark color pattern of this breed. This *light brown* allele does occur in other breeds, and can modify all other *Agouti* patterns.

The *wild* allele at the *Brown* locus causes no change – eumelanin remains black. All of the *Agouti* pattern descriptions above are undertaken as if this were the allele present. It is the original allele at this locus.

The *brown* allele is recessive and results in a medium liver shade of brown (Figure 2.15). This allele is present in African Pygmy goats in the USA, and perhaps in other breeds as well. The final brown of this allele has a somewhat redder character than the other two browns at this locus, which are more of a flat brown.

Sources of dilution in goats other than the *Brown* locus are very poorly characterized. The dilution (or intensification) of pheomelanin (tan) seems to go along two different step-wise arrangements. In one the progression is cream–gold–red. In the other the progression is cream–light tan–medium tan–dark brownish tan (Figure 2.16). In many families of goats

Figure 2.15 This African Pygmy goat has a light tan body from an *Agouti* allele, but the normally black regions have been changed to brown by the *brown* allele.

Figure 2.16 Tan can vary over a wide range of shades as are present in these Myotonic goats. Dark red (A) is at one extreme, along with a more somber mahogany shade (B). The middle range is usually either gold (C) or a flatter tan (D, also with maximum black trim for this allele). The lightest are cream or white (E).

these changes transmit as if incompletely dominant genes were responsible, with cream being one homozygote, intermediate (gold or medium tan) the heterozygote, and dark (red or dark tan) the other homozygote.

One good example of a single modifier affecting tan occurs in the Golden Guernsey goat breed. As the name suggests, these goats are ideally a rich reddish gold color. Some are much lighter, being creamier than the rich gold color desired by breeders. When pale goats are mated to dark ones, the results are generally half pale kids and half dark kids, fitting the expectation that this modification is due to a single dominant or incompletely dominant gene. Pale goat to pale goat matings do not usually occur in this breed, so it is not possible to determine what the homozygous effect of the modifier would be. Such a mating strategy, however, would solve the issue of whether the modification is dominant or incompletely dominant.

In addition, pheomelanin seems to be affected by other complicated genetic mechanisms that can either dilute or intensify. For example, the *white/tan* allele at the *Agouti* locus varies in expression from deep red to light red, gold, cream, and on to starkly white. While the locus loosely proposed for pheomelanic expression can explain some of this variation, it cannot explain all of it. It is likely that other genes at other loci, each with a small contribution, yield the final shade of pheomelanin.

A regional dilution of pheomelanin appears to be one inherent part of the original wild-type color pattern. Some populations segregate this regional dilution separately from the *wild* allele at the *Agouti* locus, indicating that there could be an independent allele at a separate locus. The details of this are uncertain.

Other types of dilution are rare or non-existent in goats, in contrast to the presence of this sort of dilution in most other species. No convincing "blue" dilution of eumelanin has been documented, nor has any evidence surfaced of a diluter that affects both eumelanin and pheomelanin.

2.4 Moon Spots

Moon spots are one of the most distinctive goat colors. These are usually round or nearly round, and are superimposed over any color of coat. Moon spots are usually pale tan, cream, or nearly white (Figure 2.17). The moon spots of kids tend to be darker (usually light brown) than those of adults. The base of hairs in moon spots is lighter than the tip, so if goats are clipped the moon spots become paler.

The genetics of moon spots has never been extensively investigated, though they do seem to be dominant. They occur in several breeds. In the USA they are most closely associated with Nubian goats and occur in Spanish, Boer, and Myotonic goats. Himba goats in Africa include a number of dramatically moon-spotted individuals. Moon spots are a breed characteristic of Sirohi goats. Goats with moon spots are highly valued in the Criolla Formoseña goat of Argentina because they are well camouflaged, thereby avoiding predators.

Moon spots vary in size and number, and extensively moon-spotted individuals can be difficult to assess as to their underlying *Agouti* pattern. Some evidence points to separate alleles for those goats with only one or a few moon spots as opposed to those that have many moon spots. On some goats with extreme manifestations of moon spots, the background color remains present only in a few isolated areas. In those goats the character of the moon spots tends to be less obviously round as the spots have merged into larger areas. Whether each of the degrees of moon spotting is due to a separate allele has never been documented.

2.5 White Spotting

White spotting, in its loosest sense, includes any pattern of either individual hairs or patches on hide or hair that fail to be populated by melanocytes. Several different white spotting

Figure 2.17 Moon spots vary from fewer and whiter (A, Himba goat) to a browner color and more numerous (B, Criolla Formoseña goat). In extreme cases they coalesce, and the background color can be difficult to determine (C, Criolla Formoseña goat).

patterns occur in goats. Each of these can be easily identified when fully expressed, and each has a specific distribution of white as the pattern progresses from minimally white to maximally white. Goats at the extreme ends of the range of variation are either least extensively white or most extensively white. The extreme goats tend to display nearly identical distributions of white and color from individual goat to individual goat. Goats in the middle ranges between the extremes tend to display more variability. A useful clue as to the specific white spotting pattern is that nearly all goats with a specific pattern retain the white areas typical of the least white goats, and tend to retain the colored areas that remain on the whitest goats. This generally stepwise progression in the development of the extent of each spotting pattern can be very helpful in teasing out the details of which specific patterns are present. Combinations of multiple patterns can occur on individual goats. In some cases it can be especially difficult to analyze the specific patterns that are present because the goats are usually extensively white.

It may well be that several of these patterns are at a single locus (the *Spotting* locus, also called the *KIT* locus) as is documented in cattle and horses. If several of these alleles are at the same locus, then not all combinations of white spotting are possible. This is true because an individual animal can have only two alleles at a single locus, and not more. The residence of all of these alleles at a single locus can also change the way they segregate through the generations. This is due to the genetic linkage arising from their residence close to one another at the single locus.

Table 2.3 The main aspects of white spotting in goats can be broken down into a few main categories of patterns.

clear crisp white areas	speckled	roan	modifications
white angora	*flowery*	*grey*	*brockle*
spotted		*pygmy agouti*	*ticking*
belted		*roan*	*smudge*
saddle		*frosted*	
blackneck			
goulet			
barbari			

The various distinct types of white spotting can be sorted into a few major groups, of which some are related to a single locus, but others are not. These include patterns with clear crisp white areas, those that are speckled, and those that are roan. Modifications of these patterns work by adding color back into the white areas in a few different ways. These are outlined in Table 2.3.

Angora White is the locus proposed as the site of a dominant allele, *white angora*, although there is at least some suspicion that the actual locus is *Spotting* (*KIT*). This allele results in a completely white coat. This allele is important in the Angora goat but is not the only source of white in that breed (Figure 2.18). Goats with *white angora* are white, and goats with the *wild* allele have color as determined by all the other loci that affect color. The *white angora* allele has not been fully characterized, but at least some evidence suggests that homozygotes are generally starkly white, while heterozygotes may have a small spot or two of color. This usually betrays itself as stripes of color in the horns, and therefore does not always affect the fleece.

Figure 2.18 The dominant *white angora* allele is responsible for the stark white of the Angora goat's fiber.

Another common form of white in the Angora breed is due to the *white/tan* allele at *Agouti*, with modification to white instead of tan. These goats can be off-white, but most are starkly white and impossible to distinguish from *white angora* goats. Having two distinct mechanisms for white in one breed makes the breeding of colored angoras very tricky indeed, because the expression of color requires that the effects of two dominant genes are disrupted. This is discussed in more detail in Chapter 4.

As with most species, goats have a recessive *spotted* allele at the *Spotting* locus. The pattern that results from this is familiar across most species. Minimally marked animals (those with the least white) usually have white on the lower legs, a star or a blaze on the face, and maybe white on the tail or tail tip. The most extensively white goats usually retain color around the eyes, and on the ears. This pattern is due to a recessive gene, and can therefore pop up as a surprise following the mating of two nonspotted goats that both carry this allele. Colored spots on these goats usually have a fairly round character. These can coalesce on the least white animals, but the round character is still usually obvious (Figure 2.19).

Belted goats have a white belt around the middle of the body (Figure 2.20). This is produced by a dominant allele, *belted*. Minimally marked *belted* goats have thin belts or incomplete belts that can be reasonably large but fail to connect over the back. White patches on both sides are common on the fairly minimally marked goats. The most extensive belts usually result in a nearly white goat with a colored head and neck. The usual pattern of

Figure 2.19 The *spotted* allele varies in expression as seen in these Myotonic goats, from fairly minimal (A), through a middle range (B), to fairly extensive (C).

Figure 2.20 Goats with the *belted* pattern have white around the middle of the body, as evident in this Myotonic goat.

color on Boer goats is a white body with red head and a blaze, and is probably a combination of *belted* with *spotted*. The *belted* allele takes most of the color off the body, while the *spotted* allele assures the blaze and white legs.

Another pattern that appears to be consistent enough to warrant some suspicion that it is caused by a specific gene adds white to the face (blaze), and around the heart girth and flank girth, but often leaves the central portion of the body pigmented. This is the *saddle* pattern which seems to pass through the generations as a dominant allele (Figure 2.21).

Another repeatable pattern is the *blackneck* pattern. It has white on the rear of the goat and pigment on the front portion. This is typical of some breeds such as the Schwarzhal and Bagot goats, but can also occur in other breeds. When the pattern occurs alone there is no white on the head. The boundary between color and white is usually crisp and sudden, and is most often found at the heart girth or somewhat behind it (Figure 2.22).

The *goulet* allele results in a white spotting pattern in Myotonic goats and some other breeds. This pattern results in irregular ragged white areas, mostly on the lower body and rear of the goat (Figure 2.23). The pattern also consistently affects the head, and can result

Figure 2.21 The *saddle* pattern has a pigmented area in the dorsal region of a belt, and generally also has white on the face, as seen in this Myotonic goat.

Figure 2.22 Valais Blackneck goats all have the *blackneck* pattern of white rear and black front.

in white around the eyes and white ears, even in fairly minimally marked goats. White in these areas is unusual for most other patterns of white spotting, and can be one indicator that this pattern is present. The *goulet* spotting pattern is likely due to a dominant allele. Minimally marked *goulet* patterned goats usually have a few irregular white spots on the middle of the sides. This is in contrast to the cleaner, single spots of minimally belted goats.

Figure 2.23 The *goulet* pattern usually has white on the ears, which is unusual for other patterns of white spotting. This family of Myotonic goats shows the progression from minimally white (A) to maximally white (D).

Figure 2.24 The *barbari* pattern nearly always has color around the eyes and nose, and most goats also have color down the topline of the back (A). The body spots vary in amount, distribution, and size, but are usually very round and obvious (B) (photos A and B by Dr. A. K. Thiruvenkadan).

White occurs on the heads of some, but not all, of these but is usually present on any grade of spotting above the minimal pattern. Maximally marked goats are mostly white, including white ears and around one or both eyes, and have residual ragged spots of color on the remaining areas.

The *barbari* pattern occurs as one variant of many in the Barbari breed of India (Figure 2.24). The most obvious manifestation of this pattern is a nearly white goat with colored lower legs, nose, eye patches, and ears, and color along the spine. The white areas contain small to medium-sized colored spots. This is likely a dominant pattern. It can appear to overlap with either *flowery* or some *brockle* phenotypes, but many of the manifestations in this breed are distinct from those and therefore this pattern is likely caused by an independent allele.

The *flowery* pattern in goats derives its name from the Florida Sevillana (Flowery Goats of Seville) breed of goats in Spain, for which this pattern is a unifying breed characteristic. The *flowery* pattern occurs in many breeds worldwide, although in most breeds it is usually just one variant among many and not a breed-wide characteristic. The *flowery* allele is dominant, and whether homozygotes are more extensively white than heterozygotes has not been established.

The *flowery* goats have small white speckles mixed into the base color (Figure 2.25). This usually takes the form of white speckles in colored areas and colored speckles in white areas where the white speckles have coalesced to form whiter areas. Goats at the lower extreme of expression usually have the speckles on the lower body, with none on the legs and head. They also commonly have a white tail and a white star on the forehead. Minimally marked goats can easily be misclassified as not having the pattern. The maximal pattern is a heavily speckled goat with darker legs, and a darker head. The bottomline of the neck and body is usually whiter than the topline of the neck and body.

The *flowery* pattern in South African and Himba goats is called "skilder" which is Afrikaans for "speckled." The *flowery* pattern is genetically distinct from the *brockle* (skilder) pattern in sheep. A pattern homologous to "skilder" sheep does occur in goats. It is a modification that adds colored spots back into white spotted areas, and is discussed in Section

Figure 2.25 The *flowery* pattern varies from minimal (A), through a medium range (B), to maximal (C). It is common on Himba goats. It nearly always involves speckles on the body and at least a little white on the crown of the head and the tip of the tail.

2.6 of this chapter. Nomenclature therefore has become somewhat hopelessly entangled for a few of these patterns in a few countries or languages.

The *belted* and *flowery* alleles are probably very closely linked. This means that the two are on the same chromosome and likely close together, making the *Spotting* (or *KIT*) locus a most likely candidate. In Himba goats most of the *flowery* goats have no other spotting pattern. In at least one family of Myotonic goats the two patterns occur together, and as a result the *belted/flowery* goats produce either nonspotted kids, or *belted/flowery* kids, but few that are only *belted* or only *flowery*.

The color terms "grey" and "roan" both refer to patterns that are mixtures of individual white and colored hairs. These patterns in goats are usually fairly uniform over the body (Figure 2.26). These patterns were long assumed to be at the *Agouti* locus due to their phenotypic similarity with patterns on sheep, but segregation data suggest that all of them are likely to be roan phenotypes that segregate at loci other than *Agouti*. This is borne out by goats with these patterns passing along two different *Agouti* alleles in addition to the roan pattern, which would be impossible if these alleles were at the *Agouti* locus. Segregation data are unavailable for some of these roan patterns, even though these are reasonably distinct on visual inspection. All of them are patterns that add individual white hairs into the background color. This is consistent with their not being at the *Agouti* locus. There is no indication that any of these goat patterns are associated with lethal traits as is the case with some roan types in sheep. Some breeds, like the Azul of Brazil, are entirely *grey* with no hint of lethality.

Figure 2.26 The Myotonic goat in A is *roan* added to a black-based *toggenburg* pattern. Myotonic goat B is a common type of *grey* that is uniform over the entire goat. The Chivo Neuquino goat C has a pattern consistent with the *agouti grey* of Pygmy goats. Nubian goat D has both *frost* and a *roan* pattern that does not affect the head and legs (photo D by A. Pattison).

The *grey* allele results in a mixture of pale and black hairs spread fairly uniformly over the entire goat, usually with somewhat paler lower legs. Some of these goats have a more speckled character to the white, rather than a distinct roan based on individual hairs. Sometimes faint stripes are present on the legs, but usually not.

The *pygmy agouti* pattern is most typical of the African Pygmy goat in the USA. It is called simply "agouti" by the breeders of this goat breed, which is an unfortunate choice of terminology because it is almost certainly not an allele at the *Agouti* locus. This allele results in a mixture of pale and black hairs, but with darker legs, back stripe, and facial markings. This allele varies in the expression of the extent of the white hairs. The lightest extreme has many white hairs, the darkest extreme has fewer. It, or a closely related pattern, is present as a variant in the Negra Serrana breed from Spain and several other breeds.

Some goats have a relatively uniform mixture of colored hairs and white hairs which is typical of *roan* patterns in many species. It is easy to confuse *roan* with *grey*, but *roan* is usually not as heavily affected with white hairs and it also generally spares the lower legs and head so that these remain dark. Minimally *roan* goats are fairly dark and easy to mis-classify, but the most extensively *roan* goats are obvious. Minimally *roan* goats are usually

Figure 2.27 The *frosted* pattern, seen on this Myotonic goat, consistently adds white to the ears, and usually to the nose and the tail.

lightest on the bottom of the neck and body, while the more extensively marked ones are usually reasonably uniform over most of the body. There are a few other candidates for *roan* patterns in goats, including one that is similar to *roan* but with a less even distribution of the white hairs. How all these somewhat similar expressions relate to one another is uncertain.

Goats with the dominant *frosted* allele have roan ears, noses, and tails (Figure 2.27). The pattern is produced from white hairs growing in among the colored ones of these areas. This varies in extent, and can be overlooked in minimally marked animals. This is a very common pattern in several breeds. It is nearly uniform in the African Pygmy breed in the USA, and is common in the Nubian (or Anglo-Nubian) to the extent that goats lacking it are somewhat unusual.

Saanen goats are white, and the few studies that have investigated the mechanisms that lead to their final white color have pointed to *white/tan* at the *Agouti* locus as an important underlying constituent. Occasional colored goats segregate from white Saanen breeding, and in the USA these are called "Sable." Most Sable goats are not very remarkable in terms of color or white pattern, making it tempting to suggest that Saanens derive their white color from the *white/tan* allele at *Agouti*. However, crossing white Saanen goats with goats of other breeds can produce goats that are white with numerous colored spots having ragged edges. This suggests that an allele that causes white spotting may also be involved. Using multiple additive mechanisms for whiteness is a common strategy for producing white animals of many species because adding distinct mechanisms together increases the odds of a uniformly white final product. The tendency for these additive mechanisms to lead to starkly white animals has the result that many white breeds hide interesting spotting or dilution mechanisms that can be brought to expression by careful breeding practices.

2.6 Modifications Adding Color into White Spotting

A few modifications of white spotting occur in goats. These are analogous to similar patterns in other species. The modifications affect the white areas of spotted animals. They have a complicated pattern of expression because the genetic instructions for these patterns can be present in nonspotted animals. However, animals lacking white areas cannot express these modifications because they can only be detected when white spotting is

Figure 2.28 The colored spots of *brockle* goats are present at birth, and usually also add somewhat larger pigmented spots to the topline. Goat A is a Myotonic goat. The *brockle* modification is a typical part of the distinctive color of Algarve goats from Portugal in photo B (photo B by L. Edmundson).

present. Consequently, these modifications can be passed along for several generations in nonspotted animals, only to be revealed when these alleles have a chance to combine with various spotting alleles.

The *brockle* or "skilder" effect (sometimes also called "calico") is due to a dominant allele that adds fairly small spots of color back into white spotted areas. These spots are generally round or oblong. They are present at birth, which distinguishes them from *ticking* (Figure 2.28). The *brockle* pattern is dramatic on heavily spotted animals, but can be somewhat subtle in minimally spotted individuals. Some animals with the *goulet* or *barbari* pattern appear to be variants of *brockle* on a largely white background, but other goats with those spotting patterns have details that suggest that *brockle* is lacking and therefore that those patterns are independent. Most *brockle* animals have spots of color along the topline, which seems to be an integral part of the action of the allele. The spots along the topline are often larger than the spots on the sides, and sometimes merge to form a nearly solid colored line down the spine. The final appearance of *brockle* goats depends very much on the extent and distribution of the white that it modifies.

Animals with *ticking* have small colored spots that grow into white spotted areas. This develops with age, so the animals are born without the tick spots and they grow in later (Figure 2.29). This usually happens by a year or so of age. The *ticking* is due to a dominant allele. It varies in extent from minimal to very heavy. In most animals *ticking* only affects the primary fibers and not the secondary ones. As a result, a goat with a heavy cashmere coat will show *ticking* in the summer, but will appear to lack it in the winter. The *ticking* can occur as a modification of any of the white spotting patterns, but is most usual in combination with *spotted* or *belted* goats. The color of the tick marks is determined by other loci, so usually the *Agouti* locus determines the color of the specific marks.

The *smudge* modification resembles *ticking* in that it includes dark hairs growing into white areas. With this modification the colored hairs occur individually rather than in spots. The final result is more roan than spotted (Figure 2.30). The extent of *smudge* varies

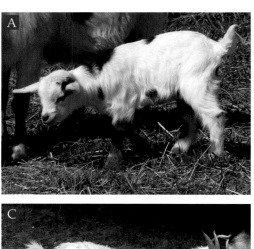

Figure 2.29 This Myotonic goat developed *ticking* as she grew older. Photo A is of the goat as a kid, photo B is the same side at maturity, and photo C is a closer view of the opposite side showing the characteristic tick spots.

considerably, and so the final appearance can be very dark but is often minimal. It is most likely a dominant allele. In many individuals the addition of colored hairs is heaviest away from the border of the white region, so that the final appearance is variable across the white spot.

Figure 2.30 The *smudge* effect is similar to *ticking*, but has a more uniform distribution of colored hairs in the white areas. It often leaves a somewhat whiter rim right around the original pigmented spots as seen in this Myotonic goat.

CHAPTER 3

Goat *Agouti* Locus Details

The *Agouti* locus patterns of goats are numerous and can easily be confusing. They have tan and black areas that are symmetrically arranged from one side of the goat to the other. The patterns are caused by many different alleles at this single locus. Fortunately, many of them share broad similarities so that they can be sorted into a relatively small number of groups of related patterns.

Observing the fine details of the *Agouti* patterns is essential for the accurate identification of these patterns, although important questions can arise with some of them because the most extensively black and the least extensively black variants are so distinct from one another. Understanding the details of the expression of this locus and its many alleles can become tedious, but the details are very helpful for some breeders with a special interest in these patterns. Undertaking a high level of detailed discussion for goats can help to establish a good understanding of general trends that illustrate how the alleles at this locus function and interact with one another. The *Agouti* locus is of special importance in both goats and sheep, with many similarities, as well as important differences, between these two species. This locus has a less complicated role in the other species.

The details of the specific arrangement of tan and black areas on the head, ears, and legs are the most consistent and useful aspects of identification for many of the *Agouti* patterns in goats. A few of those specific arrangements are repeated in several patterns, which can help to characterize otherwise similar patterns as likely to be due to different alleles. This is especially the case when the details of the patterning on the head, ears, and legs differ between two otherwise similar color patterns.

Many of the *Agouti* patterns have stripes on the head (Figure 3.1). These stripes are black in some patterns, and tan in others. The stripes are usually longitudinal and run along the head above the eye socket, or occasionally through it. Tan stripes on a black background split out into useful subtypes depending on the breadth of the stripe (wide or narrow) and its completeness from poll to muzzle. For several patterns, the tan stripes can change in the first months after birth, usually becoming narrower as black encroaches on regions that were originally tan. The result is usually then fairly stable throughout the rest of the goat's life.

Figure 3.1 The Myotonic kid in A has a white blaze, but facial stripes are evident including a light tan facial stripe above the level of the eyes, and a black stripe through the eyes. The front legs have a marking that is common to many patterns where the black is discontinuous at the knee. The Myotonic kid in B has narrow tan facial stripes above the eyes, and the black stripe on the front of the lower legs does not break at the knee. The kid in C has a more prominent tan facial stripe and only incomplete striping on the lower leg.

The other common details of patterns on the head are whether the poll is black (on tan backgrounds) or tan (on black backgrounds) and whether the muzzle is black or tan depending on the specific allele involved. The muzzle color differentiates between some patterns that are very similar in other details.

Ear color can be helpful in sorting through some of the patterns. The tops of the ears (outside) and the bottoms of the ears (inside) are the important details. The outside can be tan, tan with a central longitudinal black bar, black with a tan rim, or black. The inside of the ear is usually either completely black or completely tan.

Lower leg patterns are also helpful in distinguishing among many otherwise similar patterns. Many patterns have completely tan or completely black legs. A few lower leg patterns have specific combinations of tan and black that are consistently repeated across several different patterns. One of these is the most common and can be considered the "usual" leg pattern because it occurs in the original *bezoar* pattern as well as in several others (Figure 3.1A). The lower legs have tan posteriors and black anteriors. The front legs have black above the fronts of the knees, then a break where tan encircles the leg at the knee, followed by a black stripe down the front of the cannon region. This stripe expands to wrap around to the rear of the fetlock and extends down the front of each toe to the hoof. The rear legs have a similar marking down the front that wraps around the fetlock. However, the black does not break at the hock but instead is continuous with the black region above the hock.

A second leg pattern that is repeated in several alleles also has front legs with tan posteriors and black anteriors (Figure 3.1B). In this pattern a narrow black stripe on the front of the foreleg does not break at the knee but instead is continuous. This mark also tends not to flare out at the fetlock. In a few patterns this stripe is weaker and does not extend all the way to the hoof, varying from complete to incomplete (Figure 3.1C).

Some patterns reverse the trend of lower legs with tan posteriors and black anteriors (Figure 3.2). These patterns tend to each have a distinctive lower leg pattern that is not shared widely among several alleles, in contrast to the previous two lower leg patterns. Both the distribution and the extent of the tan areas vary from pattern to pattern. A few patterns have black lower legs but with tan regions on either the inner or outer surfaces. This contrasts to most patterns where the tan regions are either anterior or posterior.

Another detail many patterns have in common is the color of the belly. In many patterns this is tan, and usually pale tan. The extent of the tan belly varies from fairly restricted on most patterns, to some patterns where it encroaches up onto the lower body. In some patterns the belly is completely black. In others the black is somewhat restricted to the center portion of the belly with tan at the periphery.

The detail of the distribution of tan and black areas can help to distinguish patterns that are otherwise quite similar. Generally the head, ears, lower legs, and belly patterns are

Figure 3.2 Some patterns reverse the usual trend for the tan to be on the rear aspect of the legs. This Spanish goat also has obvious and well-developed, pale eye stripes down the entire length of the head.

consistent even in the most variable patterns. These details are therefore reliable indicators of the specific *Agouti* allele that is present.

The basic rule for this locus is "distinct alleles cause unique patterns." Breeders assume that the converse is also true – "each unique pattern is caused by a distinct allele." This second assumption does not always hold true because some individual alleles have variable action that can result in dramatically different patterns. Therefore the most accurate linkage of pattern to allele depends on the observations of several generations within a single family. This is especially true for the alleles that cause variable patterns. The extremes of expression of these patterns are so sufficiently distinct from one another that observers erroneously conclude that they are caused by different alleles.

The potential variation in the expression of the phenotypes of individual alleles does raise a need for some caution. Not every unique phenotype that can be observed in the field is caused by a novel allele. With that caution in mind, this discussion of the *Agouti* locus does assume that phenotypes with key distinctive differences are each probably due to individual alleles. There is a risk, however, that future evidence may prove that somewhat fewer alleles are in fact responsible for the variations observed. This discussion relies on an assumption that each distinct pattern is caused by a distinct allele in a one-to-one relationship, and in most cases that interpretation is reasonably likely to be accurate. Somewhat greater accuracy can be assumed by considering these patterns to be phenotypes and not necessarily genotypes in the narrow sense. The approach taken here may err on the side of being too optimistic in assigning allelic independence to some patterns. Nevertheless, it can still serve usefully to explore this locus and its function in goat color.

The history of assigning color patterns, and their allelic causes, to this locus has implications for evaluating the list of alleles. Patterns tend to be assigned to this locus by first noting the symmetrical distributions of tan and black areas, because no other genetic mechanism is a candidate for this outcome. Many of the patterns are then subsequently validated as being positioned at the *Agouti* locus by studies that rely on tracking the pattern through several generations. However, this second step is not true of all the patterns, and the absence of this step can raise doubts about the independence of some patterns.

Some of these patterns could indeed be nothing more than extreme expressions of other previously documented alleles that just happen to vary in the extent of black observed. An example of the general sorts of differences that can occur solely due to the extent of black comes from the *toggenburg* pattern. The least extensively black version of this pattern has the tops of the ears and the lower legs being tan. In the most extensively black version of the pattern produced by this allele a black longitudinal bar is added to the tops of the ears, and the lower leg has a black stripe that is incomplete toward the hoof. It is easy to envision these two manifestations as arising from a single pattern, and that they are nothing more than a difference in the extent of black over an otherwise similar pattern.

In contrast to the *toggenburg* example, other patterns that are superficially similar to one another have details of pigment distribution which betray consistent differences that cannot be simply related to differences solely in the extent of black. These details usually involve a reversal of a pigmentary pattern. For example, ears with a tan top and black bottom on some patterns contrast to the ears with a black top and tan bottom on others. That difference is inconsistent with being the result of only having more black pigment added. Other tell-tale differences between some of these patterns involve the belly, lower leg, muzzle, and facial striping. The list of patterns in this chapter tries to avoid the hazard of assigning unique allelic identity to different extremes of what is very likely a single pattern. This is accomplished by relying on certain specific clues that provide evidence against these patterns being nothing more than extreme expressions of a single pattern.

The various patterns in this chapter are illustrated as drawings rather than photographs to emphasize the most important details of each pattern. These drawings try to capture the extent of the black regions, and the extent, as well as the relative shade, of the tan regions in the most encountered manifestations of each pattern. The drawings employ a consistent use of ink colors:

- ⬤ Black regions represent eumelanic black
- ⚪ Tan regions represent the various shades of pheomelanic tan
- ⬤ Brown regions represent mixtures of black and tan hairs.

The numerous *Agouti* patterns can be organized into an array that starts with all tan and then progressively adds black regions, finally arriving at all black. This trajectory is not uniformly consistent and is more complex than a simple stepwise addition of black regions to a tan base. Such complexity arises because several of the patterns reverse the pigmentation of specific regions from one pattern to another. Thankfully, the *Agouti* patterns do separate themselves out into several logical groups in which the patterns are superficially similar to one another.

An overview of the goat *Agouti* locus is presented in Table 3.1. The patterns are arranged into groups based on the overall distribution of tan and black areas. The main organizing principle is that the patterns can be sorted reasonably well into these groups. The groups progress from all tan (or nearly so) toward all black (or nearly so) by sequential additions of black regions to the overall pattern. Within each group a detailed description of one main pattern is given, and then the distinctive features of each of the other patterns are briefly described. Table 3.1 provides a snapshot of the patterns and their groups that are discussed in the remainder of this chapter.

3.1 Predominantly Tan Patterns

Three patterns are predominantly tan, and they all have variable expression. Exactly where to draw the boundaries between these patterns is therefore problematic, as are the relationships between them. The "least extensively black" extreme of all of them converges on uniformly white, tan, or red. The final shade of color depends on what is happening in the "depth of tan" dimension of variation, and this is controlled at loci other than *Agouti*. The "most extensively black" end of the spectrum is usually distinctive and more easily identifiable because the black areas are somewhat different from pattern to pattern.

The *white/tan* allele (A^{Wt}) is the most dominant *Agouti* allele, and its complete dominance stands as a contrast to the more usual codominant character of the intermediate patterns (Figure 3.3). This allele causes a reasonably uniform tan coat, which can be stark white, gold, tan, or deep red depending on modifiers at other loci. Most goats with this

Table 3.1 The color patterns caused by alleles at the *Agouti* locus of goats sort themselves into seven groups. This can be a great help in identifying unique patterns that share similarities.

predominant tan	tan with black trim	tan with black periphery	black rear, tan front	black front, tan rear	black with tan stripes	nearly black
white/tan black face sable	*bezoar kolodzie luciana dark wild riedell*	*blackbelly pale group caramel serpentina moxotó curaco dinglu prairie lake lt angora dark group sajasta slovenia tan sides african tan sides corsican tan sides*	*peacock bulgaria s. carolina kahami neely prieska lorusso corsican*	*san clemente repartida moracha weyto guji*	*belly black toggenburg kravanja angel belly tan black and tan black and tan slovenia kanni adu eyebar*	*tan head formosa tan cheek near black nonagouti*

allele lack any black or pale pattern. Some do have subtle longitudinal facial stripes through the eye region. Confusingly, these stripes can be pale tan outlined with darker regions, or they can be black. Some goats have distinctly lighter heads or pale regions ventrally. A few goats with this allele have darker legs, belly, and back stripe, and a distinctive "H" shape on the head from two facial stripes connected over the bridge of the nose. The color on these is usually dark tan but can be a dark brown that is easily confused with eumelanin. When these areas are very dark the result very closely resembles the *blackbelly* pattern that is due to a different *Agouti* allele. However, on genetically *white/tan* goats these dark areas are unrelated to the *blackbelly* allele. Variation in expression of the *white/tan* allele can alter dramatically from generation to generation, so that a single genetic line can have

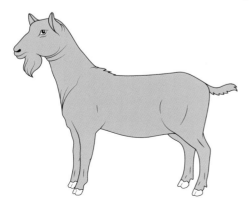

Figure 3.3 The *white/tan* allele varies in expression from white through dark red, generally with a uniform color but occasionally with light or dark highlights.

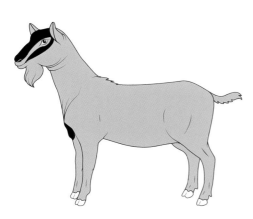

Figure 3.4 The *black face* (A^{bf}) pattern has more extensive black regions than the *white/tan* pattern. Black usually occurs over the bridge of the nose and, in extensively black examples, the goat is black on the bridge of the nose with pale stripes through the eye region and black continuing to the cheek region below the eye stripes. Some have a faint black stripe down the back, black on the knees, and thin stripes on the fronts of the legs. They also have black on the front point of the sternum. This pattern is common in some breeds, including the Nubian in the USA. The palest (lightest tan, least black) of these can be white, the darkest (red, extensive black trim) can be very dark.

great variation from individual to individual. While the depth of tan does vary widely, the amount of black is never much more than some on the head, down the dorsal midline, or at the top of the tail. The other patterns in this group include *black face* and *sable* (Figures 3.4 and 3.5).

3.2 Tan Patterns with Black Trim

The five patterns in this group share a great many similarities. They differ from the previous group in having more extensive black areas. Fine details in shade of color or distribution of black help in separating them out one from the other.

The *bezoar* (A^+) allele is the wild-type pattern for goats (Figure 3.6). This is a beautiful pattern with a shaded tan body that is darker dorsally, and paler ventrally, and has a very pale cream-colored belly. Black facial markings outline pale eye stripes. A black stripe runs down the back, and the tail usually has considerable black on its dorsal aspect. A lateral black mark is common from the spine to the shoulder, especially on males. The legs are

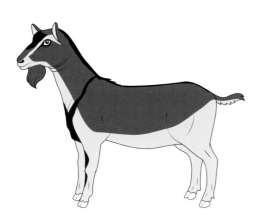

Figure 3.5 The *sable* (A^{sbl}) pattern is quite variable and is usually more extensively black than the *black face* pattern. These goats usually have a pale belly, pale legs, and pale eye stripes, while the balance of the coat on the body, neck, and head is a mixture of tan and black hairs. The *sable* pattern varies both in depth of tan and extent of black. The darkest *sable* goats are obvious; the palest can be white and can be very confusing when the darker variants are produced due to modifiers picked up from appropriate mates. The palest of these goats usually retain some darker regions as stripes through the eyes. When *sable* goats have pale tan and extensive black, they have a "striped-faced grey."

Figure 3.6 The *bezoar* pattern is the original wild-type color for goats.

Figure 3.7 The *kolodzie* (A^{kldz}) pattern is hypervariable. The middle range of expression is very similar to *bezoar* but lacks shading in tan regions. Both the depth of tan and extent of black vary. The least black ones are nearly all tan, which can produce a solid white goat when it is very pale. The most extensively black ones are a mahogany color from a mixture of black and tan hairs. Some approach the *black and tan* pattern complete with eye and leg stripes. Some have distinct pale facial stripes, others lack these.

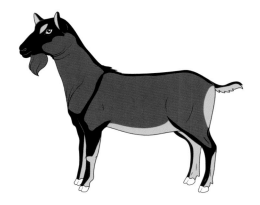

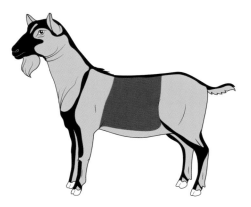

Figure 3.8 The *luciana* (A^{luc}) pattern occurs in Slovakian and French Alpine goats. It has the dramatic leg markings and pale belly of the *bezoar* pattern. The barrel and head have an uneven interplay of tan areas mixed with black that tends to vary with age, or with season. This pattern tends to be tan around the eyes, with a variable short tan bar above the eye that is much reduced from the *bezoar* marking.

distinctively marked with the usual leg markings for goats that break at the knee. Goats with the *bezoar* allele are born a bland tan shade with minor facial stripes and acquire the black pattern after a few months. Males are darker than females and tend to darken with age until they can become very extensively black. This trend of more extensively black males, and becoming more extensively black with age, is seen in several other *Agouti* locus patterns. The *bezoar* pattern is recognizable because most are a middle shade of tan with the distinctive black striping along with the nearly white belly. A few patterns, such as *dark*

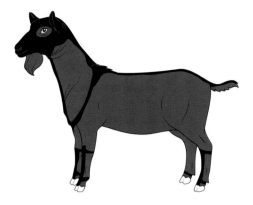

Figure 3.9 The *dark wild* (A^{dwld}) pattern is essentially a nonshaded *bezoar* with a middle range of tan. These also have a light mixture of black hairs in the tan regions, especially on the body. This pattern tends to have minimal shading and a darker belly.

Figure 3.10 The *riedell* pattern (A^{rdl}) occurs in Myotonic goats. It is variable over a narrow range that goes from a dark sooty *bezoar* pattern to a much darker more somber pattern heavily mixed with black hairs (even in females) and with little expression of the eye stripe. It lacks light shaded areas and is fairly uniform throughout. The tan belly is only rarely light cream.

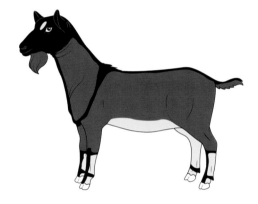

wild and *riedell*, have more uniform shading in tan regions but are otherwise quite similar to *bezoar*. The other patterns in this group are all generally darker than *bezoar*, and some are highly variable (Figures 3.7 to 3.10).

3.3 Tan Patterns with Black Periphery

The patterns in this group have more extensive black regions than those in the previous group. One pattern, *blackbelly*, illustrates the basic distribution of pigments. Seven patterns are somewhat lighter than it by virtue of having pale tan and less extensive black regions. Four patterns are darker than it because black hairs intrude on dark tan areas.

The goat pattern caused by the *blackbelly* (A^{bb}) allele is similar, but not identical, to the *badgerface* allele of sheep, although some breeders equate the two (Figure 3.11). "Bay" might be an equally good name for this pattern, and is the name generally used by Oberhasli breeders. For the Oberhasli breed this is a breed-defining pattern, although it occurs across many breeds internationally. Goats with the *blackbelly* allele are basically "tan with black edges" including black back stripe, black under the tail, black shoulder stripe, black belly, black perineum, and solid black lower legs. The head is marked with a black "H" consisting of black eye stripes that are connected at a level at or above the eyes. The "H" can expand on older males and then lose its distinctive character. The depth of tan can vary on this

Figure 3.11 The *blackbelly* pattern is common throughout the world.

pattern, but usually only from a rare gold example to the darker tan or red that is much more common. The extent of black is not highly variable, other than males being more extensively black than females and becoming increasingly so with age.

Seven pale patterns are similar to *blackbelly*. These patterns are all lighter than *blackbelly*, both in terms of generally being less extensively black as well as in consistently expressing very pale tan that is often cream or off-white (Figures 3.12 to 3.18). They are all similar, which raises a very real question about the number of truly valid *Agouti* alleles that govern the expression of groups of similar colors like this. Most of the individual patterns in this group are highly repeatable and rarely, or never, produce goats that look exactly like the other patterns. This suggests that they are each due to a distinct allele. However, some of the differences could be due to modifiers at different loci that just happen to adjust the expression of a single pattern in one direction or another. This becomes an even more significant problem in breeds of goats with fleece because the subtle differences between these then become trivial when considered against the product of the fiber that is shorn. Some patterns, such as *dinglu*, do have notable differences, such as the details of ear color, that make it unlikely that such a pattern is simply a variation from a single allele. Others, such as *serpentina* and *moxotó*, could possibly be nothing more than two extreme expressions of a single allele with respect to the extent of black. But even in this case the repeatable differences suggest that two different alleles are involved.

Figure 3.12 The *caramel* (A^{ca}) pattern of African Pygmy goats has a black blotch or stripe over the withers and down the shoulder, an extensively black face and head, and lower legs which have pale fronts and black rears. This leg pattern is the reverse of most patterns and is an important detail. The inner thigh is black. The tops of the ears are tan and the bottoms of the ears are black.

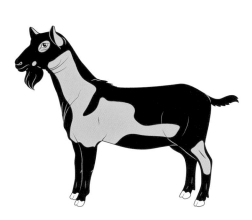

Figure 3.24 The *bulgaria* (A^{blg}) pattern has a nearly black head. The most typical examples have tan on the tops of ears, around the eyes, and a spot midway on the lower jaw. Tan is restricted to the sides of the neck and thorax, and the outside of the thigh and hock. Black goes along the shoulder as a stripe, nearly connecting the dorsal and ventral black regions. Front legs are entirely black. In the least black individuals the tan encroaches on the black shoulder stripe, and extends along the sides to include portions of the barrel. In the most extensively black individuals the tan occurs only on the sides of neck and chest, extending along the midsides to the abdomen. The tan rings around the eyes are distinctive.

Figure 3.25 The *south carolina* (A^{sca}) pattern has a dark extreme that is nearly black, with tan in the armpit, backs of thighs, base of the tail, rings around the eyes, and the tops of ears. The more extensively tan variants have more tan at the armpits, on the sides of the neck, and the backs of the thighs up to, and including, the tail base. The tops of the ears are tan, as is the base at the bottom of the ear. This pattern has a tan ring around the eyes.

Figure 3.26 The *kahami* (A^{kah}) pattern from India has black from poll to muzzle down the center of the face extending along the bottom of the chin and jaw. The tops of the ears are tan while the bottoms of the ears are black. The sides of the head, neck, and anterior chest are tan. The rear of the body is black, with tan areas on the hip and outside thigh. A black dorsal stripe is present on the neck.

The *san clemente* (A^{scl}) pattern is one of the most variable *Agouti* locus patterns (Figure 3.31).

This pattern is widespread worldwide and is routinely observed in San Clemente Island goats where it is called "buckskin." Its most recognizable form is nearly an opposite to

Figure 3.27 The *neely* (A^{nly}) pattern in Spanish goats from the USA has a black head with a tan muzzle and bars above the eyes from poll to muzzle. The ears (both top and bottom), neck, and shoulders are tan, as are the lower legs, perineum, and bottom of the tail. The body is black, as are the upper portions of the legs and the belly.

Figure 3.28 The *prieska* (A^{prsk}) pattern in Himba goats is more extensively black than most of the others in this group. Tan is restricted to the tops of the ears, sides of face, neck, and shoulder. Smaller tan regions persist in the hip region, the side of the lower thigh, and the top of the tail.

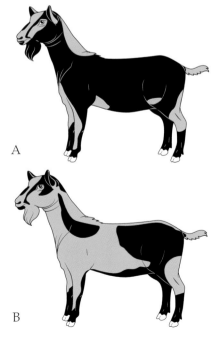

A

B

Figure 3.29 The *lorusso* (A^{lrs}) pattern varies. The most extensively black pattern (A) has tan on the tops of the ears with a central black bar, very broad tan facial stripes that tend to start above the eye and run from poll to muzzle (both of which are tan), and tan patches in front of the eyes. The topline of the neck is tan, as is the brisket from which the tan extends down to the front of the foreleg. Tan areas are behind the elbows, in front of the stifle, the entire tail, and the backs of thighs. The least black variant (B) retains the black bar on the tops of the ears, weaker facial regions, lower legs, belly, inner thighs, and rear of the body but with a tan topline. A black patch is present on the neck just below the topline.

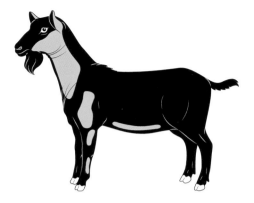

Figure 3.30 The *corsican* (A^{cors}) pattern is black with tan regions on the top of the ear, and on the head below the eye and extending to the sides of the neck. The sides of the shoulder are tan, as are the rears of the upper and lower front leg. The more extensively black variants have black on the upper ear, and tan restricted to a spot below the eye, on the side of the neck, on the shoulder, and on the inside of the upper leg.

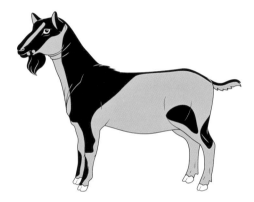

Figure 3.31 The *san clemente* pattern varies over a wide range.

the *peacock* pattern. The front half of the goat's body is black, the rear half is tan but the black regions on the topside of the tail, dorsal stripe, and lateral thigh are retained. The front legs are generally marked with usual goat leg markings that break at the knee, although completely tan lower legs do occur in the least extensively black individuals. The perineum and belly are tan, and in most cases the bottomline of the neck is also tan, with the color extending up along the back of the jaw to the base of the ear. The head is mostly black with residual pale eye stripes that are often broad as well as long. The tops of the ears are black, the bottoms of the ears are tan. Two tan spots are observed on the lower jaw behind the level of the mouth. The darkest extremes of this pattern resemble the *black and tan* pattern, with the exception that the scrotal tip is tan instead of black. The lightest extremes can resemble the *black face* pattern or can even be less black. The depth of tan observed is variable, leading to this being one of the most varied and confusing patterns at the *Agouti* locus. The palest versions can be nearly white, the darkest can be nearly black, and all except the middle range are difficult to identify. Knowing the variation across an entire family can be essential to being able to correctly identify the presence of this allele. The other three patterns in this group each have distinctive features (Figures 3.32 to 3.34).

Figure 3.32 The *repartida* (A^{rep}) pattern gets its name from a Brazilian goat breed. The front half is black, with tan on the bottoms of the ears, tan eye rings, a tan spot on the underside of the neck, and tan on the fronts of the lower front legs. The rear is tan, with black at the top of the tail, side of the hip, side of the thigh, belly, and on the rear of the lower rear leg. The center of the belly is black instead of tan. Individuals that are less black have a tan throatlatch, and the upper portion of the ear is tan with a longitudinal black bar. The *repartida* pattern varies in both depth of tan and extent of black, but the distinctive leg markings usually help to betray its presence.

Figure 3.33 The *moracha* (A^{mrch}) pattern occurs in a few Spanish breeds. The lower legs are completely black, and the belly is mostly black. The front, head, and neck are black, with a black stripe down the back. The least black variants have tan on the lower aspect of the neck and the upper legs, along with the sides of the body. The darkest ones have tan on the sides of the rear of the abdomen, and the backs of the thighs. In most cases the tan is very dark.

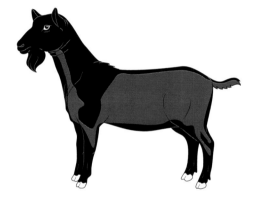

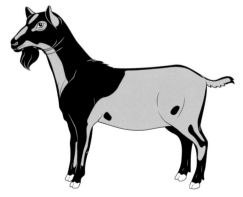

Figure 3.34 The *weyto guji* (A^{wygi}) pattern from India has a black front, which extends down the front legs but with a distinctive lateral tan stripe as well as a medial tan stripe. The tops of the ears are tan, as are the facial stripes and rings around the eyes. The ventral neck is tan, as is the belly. Black areas remain behind the elbow, in front of the stifle, and on the inner and lower portions of the rear legs. A black patch is also present laterally above the hock.

3.6 Black Patterns with Tan Stripes

The six patterns in this group have obvious facial or leg stripes, or both. Three of the patterns in this group have black bellies, and three have tan bellies. This difference further divides the group into recognizable patterns which all vary only over a limited range, and consequently are usually easy to identify. These patterns have been used as breed-specific requirements for some standardized breeds. Some of these breeds allow more than only one of the patterns because the final phenotypic appearance of several of them is so similar and therefore avoids confusion over breed identity. As is true of the other groups of patterns, most differences among these patterns lie in the details of the ears, facial stripes, and leg stripes as well as the color of the belly.

The *toggenburg* (A^{tgg}) pattern is typical of the breed of that name, although in that breed the black eumelanin of the pattern has been changed to brown by modification at another locus (Figure 3.35). In some literature this is called the *swiss marks* allele, but confusingly that name is also used for sheep with a related but different *Agouti* pattern of tan areas. When on a black background the *toggenburg* pattern is called "sundgau" in Alpine goats. This pattern is widespread worldwide. The pattern is black, with distinctive pale tan (usually cream or nearly white) lower legs and perineum, and broad pale eye stripes connecting to a pale muzzle and lower lip and extending up to the poll which is black. The tops of the ears and the throatlatch are tan. The belly is black. The most extensively black versions have a faint black stripe down the uppermost portion of the cannon region, but not extending to the fetlock. They also have a longitudinal black bar down the middle of the upper side of the ear. The tan in this pattern is almost always pale. Darker tan or red shades are very rare. This pattern keeps the tendency for pale tan in heterozygotes, so that the tan regions of the *toggenburg* pattern can be superimposed over the darker tan of patterns such as *blackbelly*

Figure 3.35 The *toggenburg* pattern is common worldwide.

Figure 3.36 The *kravanja* (A^{krv}) pattern is somewhat variable and is similar to *toggenburg* but generally more extensively black. The most extensively black expression has black on the top of the ears with a tan rim, tan above the eyes, and a tan muzzle. The lower front leg is tan except for a stripe down the front that does not extend to the foot. The perineum is tan, but not the tail. The belly is black. The lower rear leg is tan. The most extensively tan ones have tan on the bottomline of the neck, above the elbow, and a small amount in the flank.

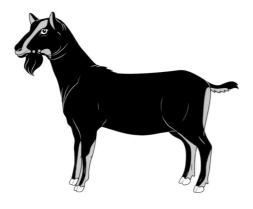

Figure 3.37 The *angel* (A^{angl}) pattern occurs in several breeds worldwide. This pattern is largely black, with pale eye stripes that are broad above the eyes, become narrower as they go to the nose, and flare out again on the upper lip. The lower lip is tan. They often have two small tan dots on the lower jaw, and a tan region on the throatlatch. The tops of the ears are black, the bottoms of the ears are tan. The perineum is tan, but the belly and udder/scrotum are black. Lower legs are pale with black stripes down the fronts. These stripes are narrow, and differ from the usual leg marks in that they are continuous on the front leg from the black body through the knee, lacking the tan break at the knee. The leg markings remain thin to the foot. The tan on these goats varies from nearly white to a light medium tan and is only rarely darker.

or others and still retain their pale character. There are two other patterns with stripes and black bellies (Figures 3.36 and 3.37).

Three patterns are largely black, have tan striping, but have obvious tan bellies. The *black and tan* (A^t) pattern is relatively common in many breeds (Figure 3.38). This pattern has a black body, black ear tops, and a distinctive array of tan areas including belly, throatlatch, eye stripes, undersides of ears, perineum, and legs with usual markings that break at the knee. On many goats the eye stripes and the leg stripes become blacker with age, being very noticeable in kids but less so in adults. In addition the eye stripes are rarely more than a small bar just above the eye, and do not extend down the entire face. The tip of the scrotum is black, and the scrotal base is tan, which helps distinguish this pattern from related ones such as *san clemente*. The tan regions vary widely across red, gold, dark tan, medium tan, and cream. The other two patterns in this group are detailed in Figures 3.39 to 3.41.

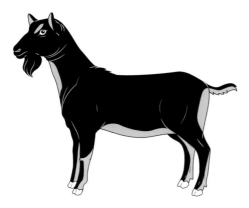

Figure 3.38 The *black and tan* pattern is very common throughout the world.

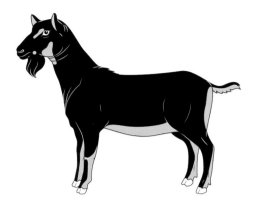

Figure 3.39 The *black and tan slovenia* (A^{ts}) pattern is similar to *black and tan* but consistently has more extensive tan areas. The throatlatch is tan, and the other tan areas are all generally more extensive than those in the *black and tan* pattern.

Figure 3.40 The *kanni adu* (A^{knd}) pattern occurs in India and Australia. It has more extensive tan than the *black and tan* pattern. The lower legs lack black markings and the belly is tan, as are the perineum and underside of the tail. There is a unique tan patch on the shoulder, as well as one on the throatlatch. A broad tan stripe goes from muzzle to poll above the eye, and the lower aspect of the lower jaw is tan. The bottoms of the ears are tan and the tops of the ears are black with a tan rim. The shade of tan varies widely from cream to dark red.

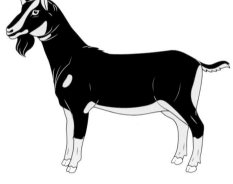

Figure 3.41 The *eyebar* (A^{ybr}) pattern shares many similarities with *black and tan* but is more extensively tan. The eye stripe is broader, the scrotum is entirely tan, and the tan belly also creeps up onto the lower sides. These tan areas often intergrade into adjacent black areas with a mixture of both pigments. The bottomline of the neck is tan, and this extends up along the jaw to the base of the ear. The lower lip is tan, and there are two small tan spots behind this on the ventral aspect of the jaw. Many goats with this pattern also have one or two small tan spots on the midline of the back. The bottoms of the ears are tan.

3.7 Black or Nearly Black Patterns

Four patterns are nearly black, and one is entirely black. These patterns lack obvious striping that is so typical of most of the other groups of patterns. These patterns are illustrated in Figures 3.42 to 3.46.

Figure 3.42 The *tan head* (A^{thd}) pattern occurs in colored Angora goats. It is largely black with a distinctly tan head that has black on the bridge of the nose as well as the bottoms of the ears. The lower legs are tan.

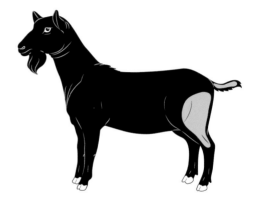

Figure 3.43 The *formosa* (A^{frm}) pattern retains tan only on the backs of the thighs and top of the tail.

Figure 3.44 The *tan cheek* (A^{tch}) pattern is widely variable but is often nearly black. The most extensively black manifestation in mature goats has tan on the tops of the ears, a tan region below the eye on the cheek, tan on the crown of the head, and tan on the upper surface of the base of the tail on the midline. Kids are born with more extensive tan areas and then rapidly become blacker, eventually achieving the usual pattern. Those with less black have tan areas that extend onto the sides of the thighs, forelegs, neck, and behind the elbow. The tan of the facial marking can extend around the eye, up to the crown, and to the upper lip. Most have a distinctive vertical poorly defined black line extending from the eye into the tan cheek. In rare cases the most extensively tan variants are barely recognizable as *tan cheek* goats because they are nearly uniform tan mixed with some black hairs, and this can include the lower legs and other regions that are generally solid black in the most typical expressions of this allele.

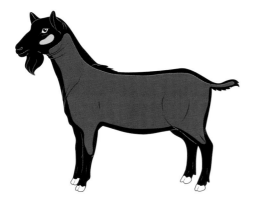

Figure 3.45 A pattern similar to *tan cheek* is *near black* (A^{nb}) which is a dark mixture of tan and black, with minor tan regions on the cheeks and lateral thigh.

Figure 3.46 The most recessive allele at *Agouti* is *nonagouti* (A^a). As the name suggests, this allele causes no tan in the coat, and the result is entirely black. It is recessive to other alleles at the locus, somewhat defying the usual pattern of codominance at this locus. A few of these goats do retain a handful of tan hairs in various locations, usually the middle of the topline, thighs, or below the eyes, but most are black throughout.

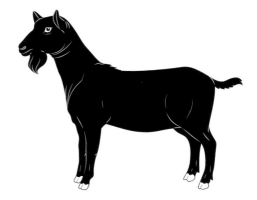

3.8 Combinations of Patterns

Goats that are heterozygous for two *Agouti* locus alleles express the tan areas of both alleles. This is an example of codominance because both alleles are expressed, and neither is hidden. The codominance of these alleles greatly helps the identification of these heterozygous goats, although some combinations are deceptive because combining the two patterns can closely resemble a third single pattern. Some *blackbelly/black and tan* heterozygotes, for example, can be mistaken for *bezoar* patterned goats, although the shading of tan areas is usually missing. Some examples, such as *peacock/san clemente* heterozygotes, can be so extensively tan as to be mistaken as *white/tan* goats. In populations with many patterns it can be tricky to accurately identify all the combinations unless the patterns of parents are known.

Heterozygotes with *nonagouti* tend to be a bit darker and more somber than are the goats that are homozygous for any of the other patterns. This tendency is not absolute but is noticed over whole population groups rather than closely related individual goat to individual goat. In addition, goats that are homozygous for any of the patterns tend to have a more extensive expression of the tan regions than those that are heterozygous.

CHAPTER 4

Putting Knowledge to Work: Goats

The genetic mechanisms behind color production are fascinating, but the real concern for most breeders is more practical: how to use genetic details to achieve their own specific goals. In some situations, the goal is to avoid the production of certain colors. In other situations, the goal is to assure the production of certain colors, or to introduce new colors into a herd. Breeders undertaking projects aimed at colors benefit from understanding the interaction of the various loci as well as the dominance relationships among the alleles. These details greatly aid breeders in coaxing the genetic pool of their goats to move in the direction they desire. A few general concepts can help breeders, and these are presented in this chapter along with the details of some specific projects that are detailed step by step so that breeders can see the concepts in action as they play out in real life.

Genetic tests for the alleles that cause variation in goat color are not widely available commercially. That may change in the future, but at least for now predicting outcomes in breeding programs requires that breeders rely on close observation of individual goats, the parents of the goats, and the offspring of the goats. Genetic tests are generally developed when researchers see the potential for enough commercial demand and a sufficient level of testing that assures them a reasonable profit. Alas, the market for color testing in goats is always likely to be fairly minimal and therefore not commercially attractive. Even if the results of genetic testing were available to goat breeders, they may well not warrant the expense. The good news is that an astute breeder with good observation skills and decent record keeping can do quite well even in the absence of test results.

Strategies for color production vary among different classes of goats. This is due to producer goals being different among the different general types of goats: dairy, short-haired meat-producing goats, cashmere-bearing goats, and the Angora goat that produces mohair.

4.1 Strategies for Color Production in Shorthaired Goats

The color of shorthaired goats is usually not a significant factor in economic value, but certain markets do place a premium (or a penalty) on certain colors of goats. For example,

Figure 4.1 In many breeds of goats the flashier color patterns can bring a premium. This heterozygous *flowery* Myotonic goat has a distinctive pattern.

some breeds such as the Swiss Alpine can vary in color, but goats marked like Toggenburgs are less favored than other colors because they can lead to confusion about breed identity. In contrast, odd or spectacularly colored Myotonic goats or Nigerian Dwarf goats in the USA are highly sought (Figure 4.1).

Breeding for consistency of *Agouti* patterns in shorthaired goats is relatively easy. Breeders can take care to incorporate patterns that are desired, and to eliminate those that are not. This is especially easy for intermediate patterns that express both tan and black areas, because nearly all heterozygotes with the intermediate patterns can be identified. Those patterns can then either be retained or eliminated depending on the breeder's goal. Modification of base color by dilution of pheomelanin, or by maximizing black areas, can also be accomplished relatively easily in shorthaired goats. As a rule, homozygotes for any of the patterns tend to have slightly more restricted black areas when compared to goats that are heterozygous for *nonagouti*. This is not consistent enough to serve as a way to reliably sort homozygotes from heterozygotes, but is consistent enough over large groups that some attention to this detail may be warranted. If a breeder desires more extensively black regions this can be accomplished by selectively retaining *nonagouti* heterozygotes. Likewise, if the goal is patterns with black regions which are less extensive, this can be accomplished by selectively retaining goats that lack *nonagouti*.

Various white spotting patterns can increase the value of goats if customers prefer uniquely marked goats. In some countries the source of this demand includes the use of hides for the manufacture of clothing and furnishings. If uniquely spotted goats are the goal, then careful selection for dramatic and crisp patterns is usually warranted. This generally means keeping the expression of the patterns neither too extensive nor too restricted, which can be a challenging tightrope to walk.

White spotting patterns can be tantalizing (or frustrating) to manage in some situations. In general, goats with combinations of multiple patterns tend to be more extensively white than are goats with only a single pattern. If clean, crisp, dramatic patterns are a desired goal, the mating of spotted goats to solid-colored mates generally tends to be a good method to practice. This reflects the fact that most of the patterns are dominant, and in most cases the

Figure 4.2 This *flowery* Himba buck is likely homozygous, and the pattern has become very white and less distinctive than the heterozygous pattern. This buck is still useful in a breeding program. He can be paired with nonspotted does for 100% *flowery* kid production.

heterozygotes have more dramatic patterns than homozygotes do (Figure 4.2). Keeping the patterns heterozygous by this strategy has the advantage of producing about 50% dramatically patterned goats. The disadvantage is that this strategy will also produce about 50% nonspotted goats. In some situations this will be enough of a detraction that the strategy of spotted to nonspotted will not work well.

The goal of targeting heterozygous phenotypes is a somewhat difficult one in any animal breeding situation. The final decisions as to which strategy to use depend on the sorts of animals available and their genotypes, and the relative demand for the less desirable phenotypes that are produced when using certain specific strategies.

The only way to assure 100% production of a desired heterozygous phenotype is to use one parent (usually the sire) that is homozygous for the desired allele. The other parent (usually the dam) is homozygous for the absence of the desired allele. The result of mating these two parental types is 100% heterozygotes, and therefore 100% the desired phenotype. Even though those two parental phenotypes may not be the actual goal, they are guaranteed to produce it in 100% of the kid crop. A drawback of this strategy is that it does not produce any of the parental genotypes. Consequently those genotypes must be constantly reintroduced into the system from outside.

Other strategies can also succeed with the goal of producing heterozygotes. This is especially practical if the less desired homozygous genotypes (and phenotypes) have other viable commercial destinations beyond retention as breeding stock. The mating of parents that are all heterozygous will produce 25% each of the homozygous types, neither of which is desirable on its own but both of which can be useful in breeding programs. The other 50% are the desired heterozygotes, and in this situation they can either be marketed, which benefits from their desirability, or be retained in the system as breeding stock.

If heterozygotes are mated to either one of the homozygous types, then the results will be 50% that homozygous type, and 50% the desired heterozygous type. The yield of

Figure 4.3 This heavily moon-spotted Criolla Formoseña kid camouflages well in brushy landscapes, and thereby avoids predators that more flashy, white-spotted kids might attract.

heterozygotes is therefore the same as with mating heterozygotes to heterozygotes, but the other 50% is now only one or the other of the homozygous (and usually extreme) phenotypes rather than both.

On a more practical level, some physical environments in which goats are produced experience high levels of solar radiation. Unpigmented skin on lips, eyes, and the perineal area can lead to skin cancers under these conditions. In such areas breeders can select for more heavily pigmented patterns, or for less extensively white goats if spotting is present. Some of the white spotting patterns tend to have white on these sensitive areas, and those patterns should probably be avoided if the goal is pigmented skin.

Other practical concerns related to other colors are also important. The region near Formosa, Argentina, is home to many goats, including the local Criolla Formoseña animals as well as imported breeds such as the Boer. This region is brushy, and has a variety of predators. Goat raisers in the region have noticed that predator losses for moon-spotted Criolla Formoseña goats are much lower than for the white-bodied Boer goats. In this situation it makes practical sense to select for goats with minimal white spotting that also have moon spotting (Figure 4.3).

Figure 4.4 This black Criollo Neuquino goat has pale grey cashmere, as is typical of black goats.

4.2 Strategies for Color Production in Cashmere Goats

Color can be important in goats used for cashmere production. White cashmere is generally the most valued color in commercial production systems. Cashmere comes from the secondary follicles, and as a result is usually much paler than the coarse hairs produced by the primary follicles. Consequently, the cashmere from black goats is usually pale grey (Figure 4.4). Indeed the Altai breed is generally black with grey cashmere, and is one of the heaviest producers of cashmere among goat breeds.

Most commercial production systems for cashmere select intensely for color. White cashmere is the target, with nearly white cashmere fiber also being acceptable in most markets. In most situations and for most breeds, these pale colors of cashmere are produced by white or grey goats. Fortunately for breeders with the goal of producing these two colors, the most common source for white goats is *white/tan* at the *Agouti* locus. Grey goats, in contrast, are produced from a handful of dominant alleles that lead to grey phenotypes that can be expressed on a wide variety of background colors. Both white and grey are produced by dominant genetic mechanisms. This situation makes selection fairly straightforward. Assuring that white or grey goats are used for breeding, especially on the male side, assures that the kid crop will be mostly white or grey.

If selection is heavy on both the male and female side, then all parental goats will be white or some variation of grey or roan. The worst case under these conditions would be that all parental goats are heterozygous for the recessive alleles at these loci. That situation would result in about 75% white or grey kids, and about 25% obviously colored kids. The assumption of uniformly heterozygous parental goats is likely untrue in most real-world situations, and consequently the production rate of colored kids in most conditions is much lower than 25%. This example illustrates the real advantages of targeting a phenotype produced by a dominant gene. Visual selection alone is sufficient to assure a generally satisfactory outcome with that goal.

Some breeders select for dark cashmere. The resultant colors are obviously "not white" selection that is targeted either in favor of or against heterozygotes for *white angora* even though this is unlikely to be fool proof in all situations. The lack of complete accuracy is likely to be more important if the goal is to select against heterozygotes, because some probably slip through with unstriped horns. The opposite situation, selecting for heterozygotes, is likely to be somewhat more successful because even though some heterozygotes are overlooked, selecting for the striped horns is likely to include very few goats that are not heterozygous.

The fact that breeding top-quality Angora goats of any color ranks right at the top of all challenging animal breeding endeavors is a useful piece of background knowledge. Keeping good mohair on Angora goats is difficult because the fleeces change with age more than those of other fiber-producing animals. Also, fleece quality (even on youngsters) seems to deteriorate more rapidly than fleece quality of other fiber species if selection pressure is relaxed even slightly. All Angora goat breeders therefore face greater challenges than most other animal breeders. Those challenges are multiplied when the goal is good-quality colored mohair because this goal is counter to centuries of selection in the opposite direction. The difficulty of breeding colored Angoras comes from adding the challenge of general Angora goat selection to the challenge of putting good color on the mohair fiber. A final contribution to the complexity of breeding top-quality colored Angora goats is that the genetic control of color in Angora goats is different from that of any other fiber-producing species. It is also different from most other types of goats.

Most Angora goats are white, black, or some shade referred to as red (Figure 4.8). Red goats are usually a medium to light shade of red. Some are born very dark red, even to the

Figure 4.8 Colored Angora goats have a wide range of fleece colors, although most are relatively pale (photo by P. Harder).

extent that it is possible to confuse them with brown goats. Most red Angora goats fade dramatically and quickly after their first shearing, their fleece color becoming a honey shade or lighter. A few red Angora goats have fleeces that darken or lighten with the season. These goats change color throughout their lives – and not always in a lighter direction. Black Angora goats usually retain black on the shorthaired areas (face, lower legs, and ears) but lighten to grey over the areas bearing fleece. Some goats remain starkly black, though. The results in the Markhoz goat of Iran suggest that deep red and true black colors of mohair are indeed possible. Some breeders note that the darkest reds and the darkest blacks tend to come out of the same families. This suggests that the modifiers for intense color are similar for both pigments. If this is the case, progress in producing one of these dark colors also assures progress with the other one.

In addition to the white, black, and red goats are goats which have patterns from *Agouti* locus alleles. These are generally classed together as "striped" because most of them have contrasting stripes of color on the face or lower legs. The most common patterns are *angel* and one similar to *eyebar*. These tend to have fairly extensive tan areas. Other patterns including *blackbelly* also occur along with a host of others. Some of the *Agouti* patterns are likely unique to the Angora goat, such as *tan head, light angora,* and *prairie lake*. An important detail on most Angora goats with *Agouti* locus patterns is that the pheomelanic areas tend to be a very dilute cream color that is nearly white as opposed to a rich dark tan color (Figure 4.9). This is a subtle but important indication that such goats are likely to have modifiers for very pale pheomelanin.

The body areas that are vitally important for *Agouti* pattern identification tend to be entirely unimportant for determining the color of mohair production. The heads and lower legs are very helpful with *Agouti* pattern details, but do not influence the color of mohair shorn from the body. Most of the more extensively black *Agouti* alleles therefore yield fleeces that are essentially black (or grey), while most of the more extensively tan patterns yield fleeces that are light tan or nearly white. Some goats have patterns with both black and tan body regions (the *peacock* pattern is one), and therefore yield fleeces with both grey and light tan fibers.

The first challenge in producing colored Angora goats is fiber quality. This challenge is best met by using top-quality Angora goats in any breeding program. Progress is always

Figure 4.9 Angora goats with intermediate *Agouti* locus alleles often have black bodies and facial stripes. The color of the facial stripes can alert breeders to the character of tan modifiers present in the goat.

Table 4.6 A cross of black goat from the red–black–brown group to registered white goat will usually result in kids that are white.

color class	Angora White	Extension	Agouti	tan modifiers	final color
white with black parent	$AW^{Wa}AW^+$	E^DE^+	A^{Wt} / any	usually pale	white

have *white angora*. Over several generations the mating of black to red within this group has accumulated the modifiers for deeper reds. This is useful for color production; however, this genetic mechanism is carefully balanced, and that balance can be thrown off by the outcross to the white goats.

The "red–black–brown" gene pool has the advantage of being relatively predictable. This is especially true when mating is restricted to goats within the group. This predictability also extends to crossing to white Angora goats, but only if the black goats from this group are the ones used for this outcrossing.

The second common group of colored Angora goats is the gene pool Coon refers to as "black," although it contains other patterns as well. The genotype of these is detailed in Table 4.7.

Goats from this "black" group are highly predictable if mating occurs within the group. The restriction of mating within the group does have the consequence that breeders are locked into the existing fleece characteristics within goats from the group. It can be slow work to improve the fleeces by mating only within the group. However, the color outcome is predictable, and this advantage has value for many breeders. Striped to striped matings will always yield striped kids, or occasionally black if *nonagouti* is present. An important detail is that the black kids from this mechanism cannot be visually distinguished from the *dominant black* ones that are typical of the "red–black–brown" group. This is one disadvantage of using the black goats from the "black" group, instead of the obviously striped ones that betray the presence of intermediate *Agouti* locus patterns. If black goats are used from this group, it is indeed important to track animals carefully so that breeders know which specific type of black goat they are dealing with.

Surprises come from the "black" group when an individual from this group is mated to either white goats or goats from the "red–black–brown" group that have either *white/tan* or *dominant black*. The crossing of goats from the two groups goes outside of the limited pool of goats whose color is determined at *Agouti*. The mating of "black" group goats to either white or to the "red–black–brown" group is nearly certain to cause results that will surprise many breeders. This results from the disruption of some of the carefully stacked genetic

Table 4.7 Diane Coon's "black" gene pool involves *Agouti* patterns, most of which have obvious stripes, but it also includes solid black goats from the *nonagouti* allele. These are all recessive to *white/tan*.

color class	Angora White	Extension	Agouti	tan modifiers	final color
striped or black	AW^+AW^+	E^+E^+	striped or A^aA^a	usually pale	striped or black

Table 4.8 Mating a "black" group goat to a registered white goat produces white kids.

color class	Angora White	Extension	Agouti	tan modifiers	final color
white with striped parent	$AW^{Wa}AW^+$	E^+E^+	A^{Wt} / other	usually pale	white

Table 4.9 The results of mating goats from the "black" group goats to goats that originated from mating white goats to "black" group goats can be surprising. They are often disappointing if the goal is a high yield of colored kids.

kid types	Angora White	Extension	Agouti	tan modifiers	final color
50% of kids	$AW^{Wa}AW^+$	E^+E^+	A^{Wt} / other	pale	white
25% of kids	AW^+AW^+	E^+E^+	A^{Wt} / other	usually pale	usually white
25% of kids	AW^+AW^+	E^+E^+	stripe / stripe	usually pale	striped

choices present in each group of colored goats. Mating a "black" group goat to a registered white goat usually results in white kids, as outlined in Table 4.8.

The kids produced from crossing "black" group goats to white goats are white from *white angora*, but they have also unfortunately picked up the *white/tan* allele (usually) and tan modifiers that are pale. The next generation of backcrossing these kids to "black" group goats (preferably striped ones) is outlined in Table 4.9.

It is easy to see that the yield of colored kids is low. This is due to introducing the *white/tan* allele along with the pale modifiers that turn a tan goat into a white one. These modifiers have come in through the registered white parent, but also from the "black" group parent. This cross is presented somewhat simplistically, and ignores the possible presence of *dominant black*, which can sometimes be introduced from a white parent although hidden in the final phenotype.

A simplistic response to this could be to resort to mating the "red – black – brown" group goats to the "black" group goats. The logic here is that color mated to color should yield color, which is not an unwarranted assumption when dealing with most genetic systems. However, the genetic systems in Angora goats are intricate and complicated, and as a consequence, surprising results occur when goats are crossed between the "black" group (generally determined by the more recessive and extensively black *Agouti* alleles) and the "red–black–brown" group (determined by *white/tan* and *dominant black*). These results can be understood by reflecting on which alleles are hidden in the various pools, but can be expressed in the crosses. It is fairly routine for breeders who cross a black and a striped goat to get quite a few white kids. The genetic machinery is complicated and is outlined in Table 4.10.

The result is that matings between colored goats have produced white kids, which comes as a disappointing surprise to most breeders that are targeting colored Angora goat production. A source of the surprise is the logical conclusion that color is recessive to white because the colored kids pop out (as surprises) from white-to-white matings in this breed. It is therefore tempting to conclude that color is a simple recessive. Unfortunately, the control of color, especially in Angora goats, involves several loci. Those loci can have complicated

Table 4.10 Mating either a black goat or a red goat from the "red–black–brown" group to a striped goat from the "black" group can often have surprising results.

	Angora White	Extension	Agouti	tan modifiers	final color
parental types for black x striped mating					
parent striped	AW^+AW^+	E^+E^+	stripe / stripe	pale	striped
parent *dominant black*	AW^+AW^+	E^DE^+	A^{Wt} / any	dark	black
kids produced					
50% black	AW^+AW^+	E^DE^+	A^{Wt} / any	pale	black
50% white	AW^+AW^+	E^+E^+	A^{Wt} / any	pale	white
parental types for red x striped mating					
parent striped	AW^+AW^+	E^+E^+	stripe / stripe	pale	striped
parent red	AW^+AW^+	E^+E^+	A^{Wt} / any	dark	red
kids produced					
100% white	AW^+AW^+	E^+E^+	A^{Wt} / any	pale	white

interactions that yield the surprising result of white kids from colored-to-colored matings even though both types of colored phenotypes are recessive to white phenotypes. Multi-locus systems are inherently complicated, especially when epistasis causes one locus to mask the effects of others.

The modifiers of the tan-based goats can be very tricky to manipulate. Some herds of colored Angora goats consistently produce colored kids. This comes from years and years of breeding goats only within the same group and concentrating on the desired dark modifiers. These individual modifiers may well be different in different herds of goats, so that even a red-to-red cross may result in a white kid if the parents are from vastly different breeding programs. On occasion even very dark red goats, when mated to striped goats, can produce white kids. This is another indication that this genetic mechanism is intricate and can be difficult to manipulate.

The genetics of colored Angora goats is complicated. Understanding the intricacies can help the breeder who is interested in crossing out of the original group to get new variants into the herd. The important issues in colored Angora goats are the presence of two whites (*white angora* and *white/tan*), a dominant black (*dominant black*), and the *Agouti* patterns, most of which have striped faces but some of which are solid white (*white/tan*) or solid black (*nonagouti*). In addition, many Angora goats have modifiers for relatively pale pheomelanin, though some escape this so that red pheomelanin is possible.

4.6.1 Truly Black and Red Angora Goats

The mohair of most colored Angoras is relatively pale compared to the completely saturated color associated with the usual pigments. While kids can be dark at birth, they often then fade as they age so that by the time of shearing, the mohair from a black kid is grey and from a red kid is tan or an almond color. This latter color is called "not quite white" by the more optimistic breeders, and while it is not truly white, it is far from an obvious and dark saturated red!

Figure 4.10 Dark mohair with fully saturated color is rare. Some black Angoras do stay nearly black (A), but most red goats fade (B) even if they are dark red as kids (photos by P. Harder).

Only rarely do colored Angora goats fail to fade (Figure 4.10). Those few can be quite useful in a breeding program if the goal is to produce the entire range of potential colors in mohair. These dark goats are likely due to somewhat complicated modifiers, although simpler systems cannot be completely dismissed in some situations.

Regardless of the actual genetic mechanism at play, selection can play a huge role in shifting the frequency of the desired dark modifiers. By retaining the blackest and reddest goats, their frequency in succeeding generations is increased and the probability of truly black and very dark red mohair therefore increases. In most situations the real goal is likely to be a fairly complete range of shades because most colored mohair is destined for use by handcrafters rather than by the mohair industry. Selecting for the darker shades is difficult in most breeders' experience, which has the consequence that actively selecting for those shades is very unlikely to eliminate the lighter grey shades, which are easier to attain. However, the color variation in the Iranian Markhoz goat does include black and dark red, which shows that success in achieving these dark colors is indeed possible.

Breeders that are especially interested in darker shades of both red and black can pay close attention to details that can help them along that path. The shade of tan that is expressed on intermediate *Agouti* patterns can be important as a clue to the character of the tan modifiers that are present and that are not expressed in black areas. A goat with very dark tan trim may not shear a fleece with red portions, but is still useful in a breeding program because it can pass those along to offspring that have picked up the *white/tan* allele from their mates (Figure 4.11).

4.6.2 Brown Angora Goats

Brown mohair is reasonably rare, despite the group name "red–black–brown." The goats designated as "brown" in that group are nearly all very dark reds. These fade to lighter colors as they age. A true brown Angora goat, likely from the *Brown* locus alleles, does occur from time to time (Figure 4.12). As kids these can be very difficult to distinguish from

Figure 4.11 This goat has the *tan head* allele. It has a fleece that is nearly black, and another important detail is the dark red of the tan regions. This goat therefore has dark modifiers for tan regions, and is useful in a breeding program that produces both black and red mohair (photo by P. Harder).

black kids because in new-borns the *dark brown* allele produces a color dark enough to be confused with black. By a few weeks of age the differences are more obvious. However it is achieved, the accurate classification of goats as to their brown status is essential if success is to be realized in a breeding program.

Dark red kids can also be confused with brown, contributing yet more problems in the accurate identification. The presence of black hairs in the tail, or along the spine, often acts as a hint that a kid is red and not brown. Truly brown goats cannot produce any black hairs, while red goats can and often do produce them.

A very useful detail that accurately establishes that a goat is brown and not dark red is the presence of any striping from *Agouti* locus patterns. When stripes are present it is nearly certain that the *white/tan* allele is absent, even though rarely such goats do have very minor leg or facial patterns. The accent stripes on goats with brown eumelanin are usually tan, and usually of a color that is distinct enough from brown that they are unmistakable.

Figure 4.12 Brown Angora goats nearly all fade to some extent. The flat brown color of the head and ears can be a useful clue to their status as *brown* goats and not red (photo by P. Harder).

Brown eumelanin tends to be a somewhat weaker pigment than black eumelanin. Fully saturated browns are fairly rare in the fibers produced by sheep and goats. Breeders can reflect on the fact that black Angoras, from whatever genetic mechanism, usually produce grey mohair, and only rarely is truly black mohair encountered. When the black is replaced by brown, the resulting fiber is that much paler and the result can be disappointing if the goal is a rich saturated brown.

If a breeder has relatively weakly pigmented, brown-based goats, these can be mated to the darkest greys, or ideally to true and fully saturated blacks if they are available. This strategy adds in dark modifiers which, when these are combined with *dark brown*, produces unique colors of mohair. Breeders have a real advantage because the *dark brown* allele is dominant. It is possible to consistently mate *dark brown* goats to darker grey or black goats over successive generations. Each of these generations produces about 50% brown and 50% black kids, and at each generational step the browns and blacks should become increasingly saturated and darker in color.

CHAPTER 5

Sheep Color

Sheep provided most of the early findings in color genetics for livestock species. The genetic control of sheep color is similar to that for goats, with a few important distinctions that make it a bit simpler to understand if goats are tackled first. The vast majority of the studies on sheep color have been accomplished using wool sheep. This is important to remember because wool sheep and hair sheep with identical color genotypes can end up looking very different from one another (Figure 5.1). Some of these differences are direct consequences of the modifications of hair coat that result in wool, and are therefore inherently part of the package that comes with all wool sheep. Other distinctions result from more optional selection pressures that have occurred over centuries and directly influence the expression of color in different types of wool sheep as well as in different individual breeds.

Figure 5.1 The Damara hair sheep in A shares a basic color genotype with the Romeldale finewool sheep in B. The differences in the hair follicles of these two types of sheep yield a very different final appearance (photo B by M. Minnich).

Figure 5.2 Solognote sheep are a pheomelanic red, but the wool they produce is nearly white. They tend to shed the wool coat in early summer, providing a dramatic contrast between the dark red coarse undercoat and the nearly white finer wool coat.

Wool fleece modifies the expression of color. Some sheep have intensely pigmented dark wool, but this situation is actually fairly rare in both standardized and landrace breeds across the world. More commonly the fleece is somewhat pale and not completely saturated with pigment due to the function of melanocytes, or a tendency to fade with weathering. The final result is that fully intense colors, such as black and dark brown, are reasonably rare in wool, and a fully intense red may actually be impossible to achieve. Dark colors do occur in some breeds though, so these are at least a possibility if breeders have these fully saturated colors as a goal.

The peculiarities of wool growth influence the final expression of color genes. A general trend is that colors in wool fibers tend to be lighter than is usual for short, coarse hair typical of hair sheep. The shorthaired regions on the face and legs of most wool sheep retain their ability to be dark. To further complicate things, the various general types of sheep fleece are different from one another biologically. These differences are important in the final expression of color genes. As a general rule, secondary fibers pick up pheomelanin very poorly, so most wool sheep in which secondary fibers predominate have very pale pheomelanic areas (Figure 5.2). Coarse and thick primary fibers pick up pheomelanin more readily, but these fibers are rare in most fleece types. Both coarse and fine fibers pick up eumelanin more easily. Primary fibers usually have a more complete saturation compared to secondary fibers.

Secondary fibers predominate in finewool breeds (Merinos and related breeds) and also in medium wool breeds (Dorset and Down breeds among others). Coarser primary fibers are more important, or at least more noticeable, in double fleeces with long coarse outer fiber and short finer inner fiber (Karakul, Navajo-Churro, Icelandic, and others). In the luster longwools (Leicester Longwool, Lincoln, Cotswold, and others) both fiber types are important. In the longwools the secondaries not only outnumber primaries, but have also undergone further modification. In longwool breeds the secondary fibers still predominate in color phenotypes. Hair sheep (Barbados Blackbelly, St. Croix, Damara, and others) reverse these relationships, so the primaries are most important and the secondary fibers are barely noticeable unless special care is taken to note them and their color.

Over the past few decades the fine details of color production and inheritance have become increasingly fascinating to a wide range of sheep producers. This is especially true among breeders producing wool fleeces for hand crafts. In light of that increased interest, it is useful to realize that many of the genetic influences on color exert very little final effect on the color of useable fleece. As a practical consequence most fleeces are white, greys of varying shades, black, or various shades of brown. This means that some of the fine details that many find fascinating actually have little practical influence on the fiber that is shorn and used. Broad overarching principles are therefore likely to be more important than fine details for those breeders interested in producing specific colors to meet market demands.

The general background of sheep domestication and selection affects color expression beyond the influences that are directly imposed by the details of fleece biology. For several millennia most wool sheep have been selected for white fleeces. A general trend in the documentation of color genetics is that most studies have been accomplished using European breeds of wool sheep. The white fleeces of most of these breeds have been achieved by selecting for completely pheomelanic pigmentation, and then subsequently diluting the pheomelanin to pale cream or white. Most white wool is the end result of this somewhat complicated cascade of events. Breeders selecting for colored wool in those breeds that are historically white therefore have to work in the opposite direction against millennia of selection, especially if the goal is a reasonably dark expression of pheomelanin. This is such a difficult task that in most situations it may actually be impossible.

The consequences of long-term selection for whiteness are most readily seen in the *Agouti* patterns in most wool sheep. In most species, the expression of *Agouti* patterns varies along two dimensions: "lightest tan to darkest tan" and "most extensively black to least extensively black." In wool breeds, each *Agouti* locus pattern tends to be very tightly clustered around an expression that consistently has very light pheomelanin as well as a minimally variable extent of black regions. Black regions are generally minimally expressed, although this varies from pattern to pattern and breed to breed. This situation contrasts to that observed in goats and hair sheep, where the range of expression of both of these dimensions (shade of tan and extent of black) varies a great deal among animals that have identical individual *Agouti* alleles. The consequence in shorthaired goats and hair sheep is that the final phenotypes of a single genotype can have a broad range of expression. The relative consistency of allelic expression in wool sheep of most breeds can be misleading when hair sheep are considered. Hair sheep vary much more in the range of expression of both shade of tan and extent of black for several single *Agouti* genotypes. These extreme expressions can easily, and mistakenly, be attributed to be the action of different alleles. The greater uniformity of expression in wool sheep can paint a relatively simplistic picture of some of the complexities encountered in other types of sheep.

Even among white wool sheep breeds, mechanisms producing their final white color vary. The whiteness of many finewool breeds relies on the addition of white spotting as

Table 5.1 Many alleles and other factors contribute directly to the basic color of a sheep.

Extension locus	*Agouti* locus	dilution	other
dominant black frequency depends on breed		*Brown* *Modified* (mioget) tan dilution	transverse stripes moon spots *aust. piebald*
wild frequency depends on breed	mostly tan or white tan with black trim grey that lightens after birth mixed tan and black interplay of tan and black mostly black	others are rare	dark points
red rare in all breeds			

well as the usual pale pheomelanic phenotype. White spotting tends to result in white hooves and unpigmented skin around the eyes and nose. The consequence of this spotting, that is somewhat hidden on a white background color, is that most colored lambs that occur as recessive phenotypes in these breeds also have white spotting. Against the colored background the spotting is more obvious. In contrast, most longwool breed associations demand full pigmentation of skin on the heads, ears, and legs on their white-fleeced sheep. As a consequence, the sheep lack white spotting and colored lambs that occur in these breeds nearly all have unspotted phenotypes.

The most important loci in determining color in most sheep breeds are *Extension*, *Agouti*, and *Brown*, as detailed in Table 5.1. Various modifications of some of the basic colors produced by these loci are obtained by other additional loci. These further modifications are very important in some breeds and are trivial in others. The frequency of alleles at the various loci is highly variable from breed to breed. Alleles that are common in one breed are often absent in others. Some alleles are frequent in several breeds, others are limited to only a few breeds. The most effective learning strategy requires breeders to appreciate the general trends across all sheep breeds, but also to become familiar with the specific situation in their own breed of interest. Table 5.1 summarizes the loci and alleles that are discussed in the remainder of this chapter.

5.1 Extension *Locus*

The *Extension* locus of sheep has two well-documented alleles: *dominant black* and *wild*, and a third rare *red* allele that provides a pheomelanic phenotype. The *dominant black* allele, as the name suggests, produces a uniformly black coat. This allele inhibits any expression of the *Agouti* locus. Several breeds have *dominant black*, and the final phenotype varies depending on the type of hair or wool (Figure 5.3). Jacob, Black Welsh Mountain, Pialdo Merino, and Hebridean sheep are all generally black due to this *dominant black* allele. In these breeds the resulting color is indeed most often black, although very dark brown can result from weathering and fading of the fiber. The short hair (head, ears, lower legs of most breeds) retains the black color expected of this allele. In hair sheep *dominant black* is

Figure 5.3 The *dominant black* allele in wool sheep often leads to a fading color that remains black only on the regions with short hair, as in the Hebridean sheep (A). Hair sheep, such as the Damara lamb (B), are usually a fully intense black.

relatively common. This is true in both African and Asian breeds, and the color is usually a strong, dark black.

The *dominant black* allele occurred historically in Shetland sheep, and some results also point to its presence in Icelandic sheep. In these and other multicolored European landraces this allele still persists in a very low frequency. Obviously, any past selection for exclusively white wool would have quickly eliminated the *dominant black* allele from a breed's genome, even if that selection only lasted for a generation or two. Even so, the *dominant black* allele has indeed persisted at very low frequencies in a number of breeds, indicating that at least a few colored sheep escaped culling. The frequency of this allele can then increase in subsequent generations if selection against colored wool is relaxed. This happened in a handful of breeds (Romney, Border Leicester, Romeldale) where *dominant black* was broadly considered not to occur. The allele must have indeed persisted in a few black sheep only to increase in frequency in the face of renewed selection favoring black-based fleeces.

The surprising presence of *dominant black* in some historically white breeds is usually attributed to recent crossbreeding or other introgression. This is no doubt true in some situations, but for several breeds in which *dominant black* is currently documented it is difficult to rationalize its introduction from the common sources of *dominant black* that are available in most countries. In the USA, for example, the usual historic sources have been Karakul, Navajo-Churro, and to some extent Jacob sheep. Those breeds each tend to leave evidence of crossbreeding in traits such as fleece character or horn form, rather than color. Most of the breeds that have recently documented the presence of *dominant black* lack any evidence of influence from those other breeds. This adds to the mystery of the source of the allele in these breeds, and leads to the suspicion that somehow the allele persisted in them and escaped all negative selection.

Figure 5.4 This Karakul was born black, and then faded to a lighter grey color. The head remains black, which is typical of *dominant black* sheep (photo by N. Irlbeck).

Some breeds have *dominant black* as the main source of black birth coats, but their fleeces then fade rapidly to various shades of grey. Karakul sheep are very good examples of this (Figure 5.4). Dark adults are very rare even though *dominant black* is an important contributor to the traditional black birth coats that are useful in the pelt production associated with this breed. The degree of fading in wool sheep depends on poorly characterized modifiers at loci other than *Extension*. The consequence of fading in some breeds is that fleece color, by itself, cannot be used to clearly establish whether *dominant black* is present or not. In some cases the fading is dramatic enough to yield very pale fleeces. Evaluating the regions with short hair (head and lower legs) can help in such animals because these retain the original black color.

Sheep with the *wild* allele at the *Extension* locus express *Agouti* locus patterns. These patterns vary, and include two that can be very confusing. One of these *Agouti* patterns is *white/tan*, and white sheep of most European breeds achieve that color by having the *wild* allele at *Extension* and *white/tan* at *Agouti,* and then rely on other loci to provide for the final white color. The second confusing pattern includes the black sheep of many breeds that have *nonagouti* at the *Agouti* locus, along with *wild* at *Extension*, which provides for the final black phenotype. This black phenotype can be identical to the one caused by *dominant black*. *Agouti* locus patterns other than these two have obviously symmetrical areas of tan (or white) and black. These are readily identifiable as *Agouti* patterns and therefore the sheep with them also have the *wild* allele at *Extension*.

A recessive *red* allele at *Extension* is very rare in sheep. The only example currently documented is the Italian Valle de Belice breed of wool sheep. Some *red* sheep are starkly white, others have tan or red markings. The *red* allele is expected to result in a completely pheomelanic coat. In wool sheep this would likely be modified to be very pale, while in hair sheep the result would be more variable, ranging from completely red or tan, all the way to white. This allele is a potentially useful source of a recessive white phenotype rather than the more usual dominant mechanism in most sheep breeds. This opens up the tantalizing possibility of white flocks that do not ever produce colored lambs.

5.2 Agouti *locus*

The *Agouti* locus is an important source of color variation in many breeds, and an increasing number of breeders are interested in the fine details of this locus and its effects. This locus has many alleles. It controls the distribution of tan and black pigment over the body in patterns that are generally symmetrical from one side of the sheep to the other. As with other species, the symmetrical tan areas of *Agouti* locus patterns are expressed as a codominant effect. An abbreviated discussion of this complicated locus is undertaken in this chapter. A more detailed approach is presented in Chapter 6.

Most studies of the *Agouti* locus in sheep have been accomplished with wool sheep. That background detail is important because coat and hair type strongly affect the expression of the *Agouti* locus patterns. Wool sheep generally have pale pheomelanin on fleeced portions of the body and also have minimally variable expression of the patterns caused by each allele. In contrast, the complete range of phenotypes on hair sheep are sometimes difficult to infer from the literature based only on wool sheep because hair sheep can have greater variability in the expression of a single allele.

The sheep *Agouti* locus is host to a very large number of different alleles and the details can lead to very long descriptions for each unique pattern. It is challenging to sort out the fine details of the *Agouti* locus in sheep, although the general trend of the dominance of tan areas can help breeders to manage these alleles in useful and practical ways. Keep in mind that tan pigment in sheep is expressed most obviously in hair sheep, and becomes pale cream or white on most wool sheep. Specific combinations of leg and facial patterns are usually reliable indicators of the *Agouti* genotype of an animal.

Despite the rich variation in *Agouti* locus alleles in sheep, the end result of most of them in wool sheep is a fairly limited range of fleece colors from white to black, through an intermediate series of grey shades of varying uniformity. In a very real sense, if colored wool production is the goal, the patterns sort out into a few categories of "white or nearly so," "light grey," "dark grey," and "black or nearly so," with a few holdouts of "variation across fleece." If black has been replaced by brown due to actions at other loci, then this range from light to dark is repeated in colors based on brown instead of black. In any event, some of the very real differences in pattern at this locus have little practical bearing on the final observable colors of fleeces.

The numerous *Agouti* patterns can be placed within broad groups that have similar distributions of tan and black areas. These groups are considered by proceeding from the most extensively tan ones down to the most extensively black ones. The intermediate patterns vary in several details of tan and black areas, some of which affect the body (and therefore the fleece) and others that do not. These six groups are:

- mostly tan or white
- tan with black trim
- grey that lightens after birth
- mixed tan and black
- interplay of tan and black
- mostly black.

Figure 5.5 The *white/tan* allele can produce entirely red hair sheep, such as this Damara (A), and is also responsible for most white European sheep such as this Merino d'Arles (B).

Mostly tan or white patterns include *white/tan* and a few others. These all tend to be very pale in most wool sheep and more obviously tan in hair sheep (Figure 5.5). Some of these have varying patterns of minor black areas remaining. In all of these patterns the fleeces are white or nearly so. The *white/tan* allele is responsible for most white sheep in European breeds. Hair sheep with these alleles have variable shades of tan, from white all the way to very dark red. In some cases a dark mahogany color is possible. The wide range of expressions in hair sheep can easily lead to confusion as to which specific allele is present.

Some patterns are mostly white or tan but have obvious black trim (Figure 5.6). These patterns tend to result in pale fleeces because the bodies are pale. Black in these patterns

Figure 5.6 The *badgerface* allele in wool sheep leads to a pale fleece, with black patterning on the legs and head. It is common in Romeldale sheep (photo by M. Minnich).

Figure 5.7 Sheep with the *grey* allele vary. The Shetland ram (A) has the typical lighter areas on muzzle, near eyes, and on undercoat of the fleece. The Shetland ewe (B) has less evidence on her face, but a fleece character typical of *grey* sheep (photo B by L. Wendleboe).

is usually on the legs, belly, face, and tail. These patterns include *badgerface* and several others. These patterns are more obviously tan in hair sheep, and account for the color of the Barbados Blackbelly breed of sheep with its tan body and dramatic pattern of black trim around the periphery.

Several patterns are mostly grey. These sheep tend to be born dark and then lighten soon after birth by the addition of white fibers (Figure 5.7). Some of these patterns retain mixtures of white and grey fibers so that the fleeces are distinctly pigmented rather than being more uniformly pale. These patterns are well-established as alleles at the *Agouti* locus, in contrast to the situation in goats where similar final colors are most definitely not at the *Agouti* locus. This group includes several individual alleles, and the first of these to be described was *grey*. The other related alleles vary in the final shade of grey or in different shades of color over the body. These patterns have been poorly explored in hair sheep, and the final appearance of these is uncertain. If the pale hairs remain truly white on a hair sheep, then the final effect is likely grey. If the pale hairs can achieve a degree of pigmentation by pheomelanin, then the final result would be a much more somber mahogany effect.

Several patterns have subtly shaded regions that vary from body region to body region. On wool sheep these usually result in fleeces having a broad range of colors but with no clear transitional boundaries between one shade of color and the next. They are especially common in longwool breeds (Figure 5.8). Many of these variable patterns, especially in wool sheep, result in a shaded grey as a final color. These fleeces can be valuable in some markets because hand spinners prize the varying colors of yarn that can be spun from a single fleece.

A few patterns have a more complicated regional distribution of tan and black areas over the body, with sharp borders between the different regions (Figure 5.9). These patterns are relatively rare in sheep, in contrast to the relatively large number of goat patterns with obvious differences in pigmentation from body region to body region.

Figure 5.8 The color pattern on this Leicester Longwool is typical of several patterns that end up with black to pale grey areas that blend into one another across the body.

Several patterns are mostly black but have distinct white or tan trim, usually as stripes on the head, legs, or belly (Figure 5.10). The result is that these fleeces are generally dark. The details that differ are usually the ear, face, and leg markings, and none of those affect the fleece. Common alleles in this group include *black and tan*. The *no pattern* or *nonagouti* allele results in a solid black sheep and is also a source of black wool for many breeds. As the name suggests, it has no tan areas.

Figure 5.9 Patterns that vary across the body manifest quite differently in wool sheep (Icelandic, A) and hair sheep (Damara, B).

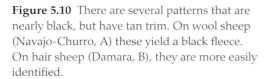

Figure 5.10 There are several patterns that are nearly black, but have tan trim. On wool sheep (Navajo-Churro, A) these yield a black fleece. On hair sheep (Damara, B), they are more easily identified.

5.3 White Sheep Versus Black Sheep

A few different genetic mechanisms can produce white sheep, and a few can also produce black sheep. These different mechanisms have consequences that go beyond the final color of the sheep. In some cases they can affect an animal's overall development and function. Individual sheep of the Portuguese Merina Branca breed (white) are generally larger than the related Merina Negra breed (black). In contrast, ewes of the Merina Negra breed are generally more reproductively prolific. This is in keeping with a trend in other species that the more dominant (and extensively pheomelanic) *Agouti* locus alleles have effects that go beyond color determination. In this case the *white/tan* allele may not be the entire story because the black color of the Merina Negra has, at least in the past, come from the *Extension* allele *dominant black*, and therefore many of the sheep may have the *white/tan* allele at *Agouti* but it can not be expressed.

The different mechanisms for production of black sheep can be important. Some breeders have noticed that the black sheep produced by the recessive *nonagouti* allele tend to keep their deep black color better than black sheep produced by *dominant black*. This has been noted in single breeds that have both mechanisms, and is a trend that may well be important for some breeders.

The presence of more than one genetic mechanism for both black wool and white wool means that breeders have potentially great challenges. In breeds where only single genetic mechanisms are in play, the results from a breeding program can be easily predicted. In other breeds, or in crossbred populations, predicting the outcomes of various mating plans can become very tricky indeed when more than one genetic mechanism can lead to a single identical phenotype. This is especially likely to happen in multicolored breeds, or flocks that have black along with other colors. The two different genetic types of black can easily persist in such populations. These two different genotypes can lead to unexpected results when they are mated together, especially if the *dominant black* sheep are only heterozygous

Figure 5.11 Transverse stripes are unusual in sheep, and have a distinct pattern which suggests important influences from messenger molecules in the embryo (photo by L. Hansen).

for that allele. Such sheep, mated to *nonagouti* black sheep, can easily produce lambs with a wide variety of *Agouti* alleles that have been hidden by the *dominant black* allele. Some breeders become concerned with having only one or the other of the two mechanisms for producing black lambs, but in reality they can both be used effectively depending on the final goals.

5.4 Transverse Stripes

An unusual phenotype of lighter transverse stripes over most of the sheep's body occurs occasionally in Southdown sheep and a few other breeds (Figure 5.11). The genetics of this unusual modification are unknown. The pattern and distribution of the stripes are reminiscent of those on some species of zebra, and they probably relate to the same basic pattern-inducing process in the embryo.

5.5 Dilution

The dilution of base colors is common in wool sheep. This is especially the case for pheomelanin. In wool fibers, fully intense and saturated pheomelanin is rare indeed. Pale pheomelanic areas likely occur due to the presence of dilution genes, but can also be due to the peculiarities of wool fiber and its relative inability to pick up pheomelanin pigmentation. Very few specific dilution genes have been characterized in sheep, but they are likely to be present and important in most breeds.

The *Albino* locus is one source of dilution in sheep, although variation at this locus is rare and has never been used to generate pale sheep across an entire breed. The *wild* allele allows for full pigmentation. An *albino* allele produces white sheep with unpigmented eyes, and has been documented in Icelandic sheep. A similar but slightly more pigmented phenotype is caused by the *albino marrabel* allele, which was found in Australian Suffolk sheep. Both of these mutations result in remarkably light final phenotypes that are not quite fully white. They also have the pale eyes typical of albinos.

The *Brown* locus has a well-documented effect on sheep pigmentation. The original *wild* allele yields black eumelanin while the *brown* allele changes all black pigment to chocolate

Figure 5.12 Moorit occurs in several breeds worldwide, and is relatively common in Shetland sheep because breeders foster a wide range of fleece colors (photo by L. Wendleboe).

brown. In some countries this is called "red." Indeed, the name for chocolate-brown wool is "moorit" in many breeds, which comes from the Icelandic word meaning "red as the moor."

The *brown* allele changes all black areas to brown, so that a sheep with any black cannot have this allele because genetically brown sheep cannot form black pigment (Figure 5.12). In keeping with most mammals, the iris of the eye is also lighter, but this is subtle in sheep because the normal iris is a yellow color rather than the deeper brown of many other species. The reflective region of the back of the eye is also changed, and this can result in a red reflection from a light beam shone onto moorit sheep. This contrasts with the usual green reflection from the eyes of *wild* genotype sheep. This test is not absolute, though, and does vary in the specific angle at which the light hits the eye as some regions of the eye of *wild* type sheep can return a red reflection. The converse is not true: the eyes of moorit sheep cannot reflect back green light.

The incidence of the *brown* allele varies across breeds. The breed standards of some breed groups specifically emphasize a bluish color to the face and legs. This is especially true of several of the longwool breeds. Among longwools the moorit color is rare because of long selection against any color that is not black-based. This is due to the moorit geno-type not yielding that final desired blue color which depends on black skin under a white hair coat. While moorit does occur rarely in Lincoln and Wensleydale sheep, it is unusual and typically needs targeted selection to bring it forward to expression. Nancy Irlbeck has accomplished this in the Wensleydale breed. It is possible to increase numbers of sheep with moorit color once they are finally located within a breed, although it does take atten-tion to detail. Strategies for accomplishing this are outlined in Chapter 7.

Other breed groups, especially the finewools, tend to have a higher incidence of moorit cropping up as unexpected colored lambs. A potential reason for this is that white finewool sheep have undergone centuries of stringent selection against any stray colored fibers. If those fibers are black, they are more easily detected than would be the case if they are moorit or some other derivative of it. Animals with stray colored fibers have traditionally been culled and not used for reproduction. However, moorit-based sheep can more easily escape that fate than black-based sheep. This seems to be a somewhat trivial difference in selection pressure, but when multiplied over large populations and over several generations it has a significant influence on the frequency of the underlying alleles.

Figure 5.13 Mioget is distinctly lighter than moorit and is common in Shetland sheep (photo by L. Wendleboe).

A dilute modification of eumelanin has long been known in Shetland sheep and is recognized as a different color. The most dramatic expression of this is a honey-colored fleece, called "mioget" in the Shetland nomenclature for color (Figure 5.13). The genetics of this effect have been investigated by Linda Wendleboe, and her findings are consistent with a recessive modifier, *modified* (*Mod^{mod}*), at the *Modified* locus. The interactions of this locus with eumelanin are complex. It was possible to trace the details of this color in Shetland sheep because in some family lines of that breed the eumelanic phenotypes do not fade very much. This allows the effect of *modified* to be noticed, where it might not be so clearly evident on sheep of other breeds with fleeces that fade more dramatically due to other poorly characterized modifiers. The final effect of this dilution would be minimal in such situations.

The various combinations at the *Modified* and *Brown* loci give a wide range of final colors because the alleles at each of these two loci are incompletely dominant at least some of the time. Various combinations are outlined in Table 5.2. Sheep that are homozygous for the recessive alleles at both loci (*modified* and *brown*, or *Mod^{mod}Mod^{mod}*, *B^bB^b* as a genotype)

Table 5.2 Color phenotypes of varying genotypes at the *Brown* and *Modified* loci. The variable dominance at both loci provides for a range of colors both within and among the genotypes.

Brown genotype	*Modified* genotype	color phenotype
B^+B^+	Mod^+Mod^+	black
	Mod^+Mod^{mod}	off black
	$Mod^{mod}Mod^{mod}$	cool pewter
B^+B^b	Mod^+Mod^+	warm black or black
	Mod^+Mod^{mod}	dark brown, warm black, sometimes black
	$Mod^{mod}Mod^{mod}$	warm pewter
B^bB^b	Mod^+Mod^+	brown (moorit)
	Mod^+Mod^{mod}	fawn or moorit
	$Mod^{mod}Mod^{mod}$	honey (mioget)

Figure 5.14 Pewter Shetland sheep yield a blue-grey fleece with a unique character (photo by L. Wendleboe).

are lightened to the light yellowish or honey mioget color. Sheep homozygous for the dominant alleles *wild* and *wild* are black (*Mod⁺Mod⁺*, *B⁺B⁺*). The intervening combinations produce colors intermediate to these extremes. The result is a two-locus system that allows for several subtly different fleece colors controlled by only two loci. This is because both loci exhibit imperfect dominance between their two alleles. This system is most noticeable in sheep where colors fade minimally. Close inspection of areas with short hair (face, legs) reveals some level of action, although in those areas the effects can be subtle.

While the differences between black, moorit, and mioget wool are substantial, the intermediate colors are more subtle and range from pewter to a very dark brown that is nearly black (Figure 5.14). These fleeces can be useful for hand spinners because the unique color stands out well in single fleeces. In pooled fleeces, such as is necessary for commercial production, the intermediates would be lost in a general "grey" or "brown-black" mixture.

The various degrees of final color can have interesting and useful consequences. Some breeders, for example, have noted that sheep homozygous for *black* tend to produce more blue shades of final color, while those that are heterozygous for *brown* tend to produce somewhat more brown shades. Breeders paying close attention to these details can achieve unique fleece colors. Some of those will meet high demand from hand spinners. The effects of *modified* on hair sheep have never been documented, and are likely to be minimal.

Rarely do sheep have a *dark brown* variant that is distinct from the usual moorit brown caused by the *Brown* locus (Figure 5.15). The locus at which this variant occurs has never

Figure 5.15 The *dark brown* allele is incompletely dominant. The front lamb lacks the allele and is black, the middle lamb is heterozygous, and the back lamb is homozygous (photo by R. Lundie).

been determined, but it is not likely to be at *Brown*. The *dark brown* phenotype is due to a dominant allele. The heterozygotes are born nearly black, and then become obviously brown by the time of weaning. This change includes the face and legs as well as woolled portions. Homozygotes are a much lighter color that tends to be uneven over the body. This is a different final color than the *dark brown* of goats, which does not vary over the body, and suggests that the genetic mechanism is not the same.

Another candidate for a specific dilution allele occurs in Navajo-Churro sheep (Figure 5.16). This breed has an essential role in the production of the unique textiles produced by both Navajo and Hispanic artisans due to its characteristic fiber quality along with its wide range of colors. Among these many colors are several based on *dominant black* at *Extension*, a host of *Agouti* alleles, and also both moorit and mioget. One phenotype is a distinctive blue grey. This color is referred to variably as "Navajo Sheep Project blue (NSP blue)," "dilute blue," or "Navajo blue." This last name is descriptive and specific, capturing both the unique color and its origin. Navajo blue is likely to be a derivative of *dominant black*. The lambs are born black, but then at about six months the fleece develops a uniformly pearl grey color. The face and legs remain black. The blue color, once attained, does not change from year to year. The belly is blue, in contrast to the usual black sheep in the breed that fade with age and have black bellies.

Figure 5.16 Navajo blue fleeces fade to a blue grey early in life (photo by C. Dvergsten).

Navajo blue sheep have an unusually high luster. This is not only true of the fleece, but also the shorthaired face and legs. The luster is most easily noted on sheep immediately following shearing, but can even be noted on lambs if they are inspected carefully. The Navajo-Churro is a double-coated breed, and high degrees of luster are not found in most sheep. Close studies of the transmission of this color over generations have not been done, and are somewhat thwarted by the wide range of colors in the breed and the likelihood that this color is the result of a very specific combination of alleles at different loci that fail to yield a distinctive phenotype when matched up with the many other possible combinations of alleles present in this variable breed. Navajo blue is a distinctive color phenotype that is generally either completely present or completely absent. This strongly suggests that it is due to a single allele that modifies black, and most likely only *dominant black*.

Dilution of pheomelanin is important and common in sheep. The locus or loci causing the elimination of pheomelanin from most wool sheep has never been well documented. Part of the dilution process in wool is due to the biology of the wool fiber itself – secondary and fine primary fibers simply do not pick up pheomelanin very effectively. It is also likely that single genes of major action play a role in the removal of pheomelanin. Hair sheep vary widely in the depth of saturation of pheomelanin, so clearly other factors are affecting this besides the peculiarities of wool (Figure 5.17). It may well be easier to study the genetic

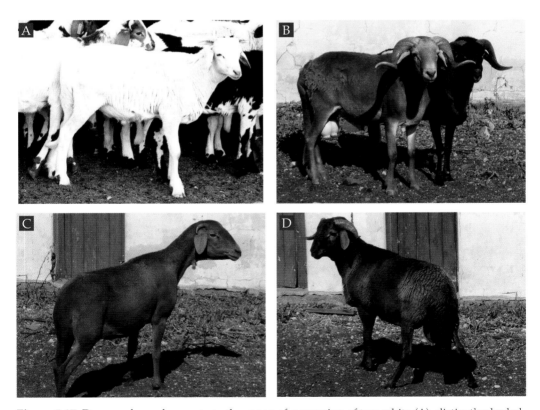

Figure 5.17 Damara sheep demonstrate the range of expression of tan: white (A), distinctly shaded light red (B), red (C), and mahogany with considerable black in it (D).

Figure 5.18 Moon spots stand out distinctly in hair sheep (photo by D. duToit).

mechanisms behind whiteness in hair sheep than wool sheep, largely because hair sheep have not undergone intense selection for whiteness. White sheep do occur in some hair sheep breeds (Damara, St. Croix) and might prove to be a useful starting point.

Other types of dilution include Jacob sheep that are lighter than the breed's *dominant* black. These are called "lilac" and include moorit, mioget, pewter, *Agouti* patterns with grey fleeces, or other undocumented mechanisms. Sometimes Finnsheep are born with pale red wool and darker red legs and heads. A second Finnsheep color is born with light grey wool and black legs and head that fades to a nearly white fleece with very pale legs and head.

5.6 *Moon Spots*

Moon spots are round tan spots that occur over any background color as determined by the *Agouti* locus. These are common in goats but are very rare in sheep. In wool sheep they are likely to be completely overlooked, while in hair sheep they can be as distinct and obvious as they are in goats (Figure 5.18). In goats they are likely dominant, and this is therefore likely in sheep as well.

5.7 *Australian Piebald*

Australian researcher Scott Dolling encountered a phenotype of isolated large black spots on otherwise white sheep (Figure 5.19). Further investigation revealed that these black spots can be superimposed over any *Agouti* pattern. In Australia, the background pattern is almost invariably white (from *white/tan*), but the same black spotting can occur on *badgerface* or any other *Agouti* locus pattern. The spotting is due to an incompletely penetrant recessive allele *piebald* at the *Australian piebald* locus. This should never be confused with the *spotted* allele at the *Spotting* locus because the mechanism of action is completely different. In this case, the mechanism is a black spot lying over any *Agouti* phenotype, and not a residual black spot on an otherwise white-spotted phenotype.

Figure 5.19 The *piebald* allele is the cause of random black spots that occur on Australian Merino sheep (photo by S. Dolling).

Confusion between these two biologically distinct mechanisms is easy when the *Agouti* phenotype is from the *white/tan* allele. The *piebald* allele is problematic for wool producers because of its irregular genetic transmission, and also because the final phenotype is neither white nor colored.

The spots caused by the *piebald* allele do not consistently appear on specific regions of the body, but are distributed randomly. Sheep usually have just a single spot. These spots can present a real challenge to sheep breeders. Some breed associations take a firm stance against random black spots on otherwise white sheep, which is wise because this allele is one likely source. Some sheep breeders think that a random black spot on a white sheep might indicate a sheep that carries recessive color, but this is not usually the case. Sheep with these spots offer little to the breeder interested in producing colored sheep, and at the same time are a significant threat to the breeder interested in clean white wool on white sheep. This allele therefore does not really serve the goals of the vast majority of sheep breeders.

5.8 Sur

Two different recessive alleles, at two different loci, control the occurrence of pale tips on birth coat hairs of Karakul sheep. These pale tips occur on various background colors including the common *dominant black* of that breed (Figure 5.20). This indicates that the pale tips are not an *Agouti* phenomenon. The two different types of this color are called Bukhara sur and Surkhandarya sur. They are both very similar in phenotype, but they each occur in two distinct geographic regions of Central Asia and are therefore considered to be separate. The effects of the pale tips on the phenotypes of older sheep past the birth coat stage are undetermined, and the effects on hair sheep are unknown. These same pale tips can be noted on some black Angora goat kids at birth, most of which are *dominant black* at the *Extension* locus. The pale tips can be widespread over a lamb, and can be short or relatively long. On many lambs the color can only be appreciated by parting the hairs. This is a valuable color in Karakul sheep because it produces unique lamb pelts. Sur of both types is reportedly due to recessive genes, and can modify any base color. The segregation data are more complicated than for a simple recessive phenotype, though, and suggest at least some polygenic influence.

Figure 5.20 Sur color is pale tips on a dark birth coat of Karakul lambs. The result is subtle but of great use in the pelt industry. The color is subtle from a distance (A) but more noticeable by parting the fibers of the birth coat (B) (photos by V. Tymoshchuk).

5.9 Dark Points on Down Breeds

The dark head and legs of breeds such as Suffolk, Hampshire, and Southdown have proven enigmatic for several investigators (Figure 5.21). Several theories have been put forward to explain the genetics behind these. Many of the theories involve mutations at the *Extension*

Figure 5.21 The dark shorthaired regions of the Down breeds are confusing. Fully dark in some breeds (Shropshire, A), they become speckled in other breeds and crossbreds (Navajo-Churro, B; Criolla Formoseña, C).

locus. Several problems present themselves with just about any theory, and this color remains largely unexplained.

The usual phenotype when this phenomenon is fully expressed is a lamb with a dark birth coat. The fleece then gets lighter as the lamb grows. Most fleeces end up being white or nearly white. The pigmented areas on the face and legs remain dark and do not fade. These dark areas vary from breed to breed in color as well as in the extent of coverage. In most purebreds of the Down breeds (Suffolk, Shropshire, Hampshire, and others) the head and legs remain black or nearly so, with a solid uniform color and no mottling. This is present on both the head and legs. Crossbreds produced between these breeds and fully white breeds, especially the finewools, tend to have heads and legs that are mottled to varying degrees with white and dark patches. These dark patches can vary in color from black, through brown, and to yellowish tan. The dark areas tend to center on the eyes, ears, nose, and lower leg near the foot.

All individuals of some breeds, such as the Beulah Speckled Face, have mottled faces as a breed characteristic. This likely indicates that whatever the genetic mechanism for colored points, it appears to be polygenic and also can respond to selection so that it manifests consistently at different intermediate degrees rather than "all or none." Another confounding phenomenon is that the final appearance of some sheep is remarkably similar to the phenotypes produced by various spotting alleles such as *turkish* at the *Pigmented head* locus. This makes final assessment as to the specific genetic mechanisms involved quite difficult if the phenotypes of just a few individual sheep are all that are available for observation. Observations across an entire population are more likely to sort through some of the potential confusion.

Another confounding phenomenon with this color and its genetic cause comes from the black lambs that are occasionally produced by Suffolk sheep. Suffolk sheep generally have white wool, and the appearance of black lambs from white-fleeced but black-pointed parents suggests the usual recessive mechanism at *Agouti*. However, in several instances these lambs have then gone on to produce lambs as if the black were due to a dominant allele. This makes for a very complicated situation, and one that is difficult to unravel as to the exact genetic mechanisms at play.

5.10 White Spotting

Sheep have a great number of different and distinct white spotting patterns, each caused by a separate genetic mechanism. Each spotting pattern can be superimposed over any background color. The identification of white spotting patterns can be daunting, especially when considering animals that have combinations of two or more of the white spotting patterns. The patterns each show up dramatically in hair sheep, where they are more easily identified than they are in wool sheep because the fleece often obscures fine distinctions that separate the patterns.

The orderly progression of the pattern of each different white spotting allele can help to determine which allele is present. At the minimal (least white) and maximal (most white) extremes, each pattern converges on a very repeatable distribution of white and color. The middle portion of the range of expression can be more confusing, but it is rare for an animal

Table 5.3 The general categories of white spotting each have several different specific patterns.

clear crisp white areas	speckled	roan	modifications
spotted	*flowery*	*roan*	*brockle*
bizet		*snowy*	*ticking*
belted		*lethal roan*	*smudge*
persian		*pedi*	
turkish			
afghan lethal			
white tail tip			
wading			

with a specific spotting allele to have color on a body area which tends to be white on those animals with minimal expression of the pattern. It is likewise rare for an animal with a specific spotting allele to have white on a body area which tends to be colored in cases of maximal expression of the pattern. It is somewhat useful to imagine that each pattern progressively expands from the minimal expression to the maximal expression. In the vast majority of cases this expansion is orderly and predictable. Each pattern is therefore usually identifiable.

The white spotting patterns of sheep can be organised into a few different categories. Some of them have clear, crisp white areas, others are speckled, and a few are roan. Some modifications that put color back into the white areas are also important. These are outlined in Table 5.3.

Sheep have three alleles documented to be at the *Spotting* locus (Figure 5.22), although several other patterns are also candidates. The *wild* allele results in a nonspotted phenotype. The most common mutant allele causing spots is *spotted*, which is a recessive allele. When minimally expressed the white tends to be on the tip of the tail, head (star or blaze), feet, and lower legs. In the middle range of expression the white is widespread over the body, and usually the colored areas are somewhat circular. Maximally spotted animals usually retain color on the ears and around the eyes. Another allele, *bizet*, has been proposed as the

Figure 5.22 Bizet sheep (A) have a spotting allele that puts white on the face, legs, and tail. This allele is probably widespread in many breeds. A second common allele, *spotted*, allows for more white regions over the sheep (Icelandic Sheep, B).

Figure 5.23 The *belted* pattern is repeatable from generation to generation in Damara sheep (photo by D. duToit).

cause of low grades of spotting (on the tail tip and head). This is the pattern of minor white marks on the Bizet breed of sheep in France. This allele is likely partially dominant, so that some but not all heterozygotes have some evidence of spotting.

A multigeneration experiment, using Icelandic sheep, selected piebald sheep to have maximally expressed white areas. The end result included sheep that were nearly entirely white. The wool on these sheep was pristinely bright white, which is typical of white wool that results from a mechanism involving white spotting because the white areas have no melanocytes that can cause even small amounts of pigment to be placed into wool fibers. This contrasts with the usual "white" wool from the *white/tan* allele at *Agouti* and which usually has a somewhat ivory color rather than stark bright white.

The genetics of the *belted* pattern in sheep have never been investigated. Belts are dominant in goats and cattle, and are therefore likely to be the same in sheep (Figure 5.23). The *belted* sheep have white encircling the body. Minimally *belted* animals have white on the midsides, and maximally *belted* animals can have entirely white bodies.

The *Pigmented head* locus is important in hair sheep and in a few wool sheep breeds that hail from widely scattered countries (Figure 5.24). A handful of different alleles are documented. These alleles vary in the extent of white that they produce. The phenotypes of these tend to overlap to some extent, so determining which specific allele is present can be challenging.

The pattern that has the most extensive pigment is caused by the dominant *persian* allele. This allele is nearly uniform throughout the Blackheaded Persian breed. The pattern, when most extensive, has a completely colored head that can extend onto the neck, and a white body. The extensive pattern is most common in homozygotes, and also in breeds such as the Blackheaded Persian that are selected for this maximal expression. Less extensively white manifestations of the pattern are common on heterozygotes. Such sheep usually are mostly pigmented, but with white spots on the sides. These sheep tend to have a completely colored head and legs, and the pattern tends to be roughly symmetrical. This helps to separate it from other alleles, such as *spotted*, that tend to put white areas on the head and legs.

Figure 5.24 The Damara sheep A and B have the *persian* allele. A is homozygous condition, B is heterozygous. The pattern on Walliser Scharznäse (C) and Brillenschaf (D) is also likely at this locus. The *turkish* allele causes a pattern that is extensively white (Damara, E).

The *turkish* allele is dominant, and generally causes more extensive white than the *persian* allele. The most extensive pattern is usually a white sheep with color around the eyes, nose, ears, lower legs, and middle of the belly. Less extensively white animals have colored spots on the body, but usually do have white on the head which contrasts with the *persian* pattern. This pattern is common in some Turkish breeds. It is also the likely mechanism behind the color of several European breeds such as Brillenschaf, Walliser Schwarznäse, and several Churro breeds from the Iberian peninsula. Selection keeps the fleeced regions white, although lambs vary and some tend to end up with colored body

Figure 5.25 The *white tail tip* of Damaras is a very consistent and repeatable pattern.

spots. These colored areas are somewhat more common on the rear end of the sheep than elsewhere. Wool sheep with this allele have starkly white fleeces, in contrast to the pale ivory color of most breeds that are white from the *Agouti* locus mechanism. It is therefore one of the mechanisms used for white wool production in some breeds.

The most dominant allele at the *Pigmented head* locus is *afghan lethal*. The phenotype is more extensively white than *turkish* with colored spots around the eyes, on the nose, on the ear tips, and some on the legs. Homozygous lambs are entirely white and die soon after birth. Only heterozygous sheep survive to reproduce, so this allele has not been widely used to generate white sheep. Although the sheep with this allele have starkly white fleeces they have the major drawback of not consistently producing white lambs because they are all heterozygous.

Especially in the Damara breed, many sheep are completely colored but with a *white tail tip* (Figure 5.25). The white can extend to include most of the tail. The consistent expression of this pattern suggests that this might be a separate allele for a white-tipped tail. While low grades of *spotting* can also cause a white tip to the tail, those animals usually also have white on the head and lower legs which this pattern lacks.

A few sheep have distinctive white legs and a white belly. This pattern is similar to the "wading" pattern of cattle, and is likely to be caused by a distinct *wading* allele (Figure 5.26). Most of these sheep have a sharp demarcation between the colored upper body and the white lower body and legs.

Sheep have a pattern of white spotting that has some resemblance to the *flowery* pattern of goats (Figure 5.27). This pattern consists of small flecks of white in colored areas, and small flecks of color in white areas. The pattern varies in extent. Minimally marked animals usually have white on the lower body. Maximally marked animals usually have color remaining on the head, legs, and rear of the body. This is dramatic and obvious in hair sheep, but can be subtle and missed in wool sheep unless they are shorn.

Roan is a general term that describes mixtures of colored and white hairs. Roan patterns in sheep are variable, and some of them include roan areas that are scattered over the body in distinct patterns. Some of these patterns are repeatable and are therefore likely due to

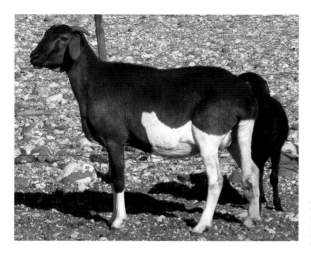

Figure 5.26 Damara sheep with white legs, tail, and lower body have the *wading* pattern.

distinct alleles. Roan patterns of goats and cattle are not caused by *Agouti* alleles, but in sheep, *grey* and related alleles at the *Agouti* locus do result in phenotypes that are mixtures of white and colored fibers. This is roan by definition. A few other roan patterns occur in sheep, generally in hair sheep, and these are more typical of the roan patterns in other species and are unlike the various *grey* patterns of sheep. For those patterns especially, it is likely that they are due to alleles homologous to the alleles in other species. A few of these roan patterns are reasonably repeatable, and stand out as warranting individual description.

"Snowy" has been proposed by Roger Lundie to designate one of the roan patterns seen in Damara sheep. The roan areas affect the upper body much as if the animal had been in a snowstorm (Figure 5.28).

Less distinctive roan patterns also occur in sheep. These tend to be much more uniform than *snowy* sheep. In some, the head and legs are darker than the body. This is consistent with roan patterns in several other species. Some roan sheep have pale heads and legs, and in some cases these are nearly white. It is likely that these two patterns are genetically different (Figure 5.29).

Figure 5.27 The *flowery* pattern of sheep is common in the Damara breed. The pattern is more ventral and anterior than the similar pattern of goats.

Figure 5.28 The *snowy* pattern of Damara sheep has roan areas over the top of the sheep.

The *lethal roan* of Karakul sheep is relatively uniform over the entire lamb. This pattern is useful for the production of silver-grey lamb skins when it is present on a black background color (Figure 5.30). Heterozygous lambs are a medium blue-grey color. Homozygous lambs are very pale roan, but die soon after birth from faulty intestinal nerves related to the pale color. This allele has been used to generate grey pelts for the fur trade based on Karakul lambs. The lethal character of the allele means that breeders can never achieve 100% production of the desired grey lambs.

Especially in hair sheep, some roan animals have distinct patches of non-roan areas scattered over the more usual and uniform roan areas. This is very similar to the *grey* pattern in alpacas, which is a dominant allele in that species. The extent of the non-roan patches varies, and their distribution seems to be fairly random. This is somewhat similar to the Pedi strain of Nguni cattle, but aside from that is like no other pattern of goats or cattle.

A pattern called "raindrops" or "flowing droplets" by Roger Lundie is a generally white sheep with small, oblong flecks of color. These are more concentrated on the topline of the

Figure 5.29 Some roan Damara sheep have a variable pattern.

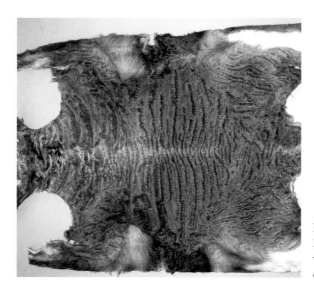

Figure 5.30 The *lethal roan* allele of Karakul sheep produces the grey color valued in the pelt trade (photo by D. duToit).

sheep, with more obvious spots of color on the lower legs and rear, and around eyes and ears. This might well be a distinct pattern, but there could also be a combination of multiple other patterns leading to this final appearance.

5.11 Modifications of White Spotting

Sheep have a few consistent modifications that affect white spotting. These add color back into the white areas that are determined by the various spotting alleles. These modifications seem to be able to do this on any of the white spotting patterns, and it does not matter which specific white spotting allele is present. The genetic instructions for these modifications can be present in nonspotted sheep, but on such sheep they are obviously not evident because white spots are necessary for these modifications to be expressed.

A very distinctive pattern called *brockle* ("skilder" in Afrikaans) puts medium to small round spots of color back into white areas (Figure 5.31). In goats and cattle this is almost certainly due to a dominant allele, and it is likely to be the same in sheep. The combination of *brockle* and *persian* leads to a very dramatically marked sheep. This is the genetic formula for the Skilder Persian color pattern. Animals with *brockle* commonly have a higher density of colored spots along the topline. The spots along the topline are also usually larger than those elsewhere. This distribution and character of spots seems to be an integral part of the action of the allele, and is a feature shared across sheep, goats, and cattle.

Another modification of white spotting is ticking (Figure 5.32). This is due to a dominant allele. Small (or more rarely, large) round to oval spots of color grow back into white areas. This change occurs after birth, and can be delayed, occurring at around a year of age. In most cases the ticking spots affect primary follicles more than secondaries and as a result they can be missed in some seasons of the year as well as on some hair or wool types of sheep. The result of ticking on Jacob sheep is called "freckling," but is due to the same *ticking* allele as in other breeds. The ticking in Jacob sheep varies from absent (most usual),

a final white phenotype. These breeds tend to produce *Agouti* patterns on which the tan is fairly obvious, especially on the legs and head. The colored sheep of these breeds also tend to have white spots, which are much more noticeable on a colored background than on the pale background afforded by A^{Wt}.

Other breeds, most notably the luster longwools, have been selected to have white wool but fully pigmented skin. The desired final appearance of these breeds features pigmented skin on the nose and around the eyes and ears, and black hooves. They therefore lack *spotting*. The colored lambs from these breeds usually have tan-based areas that are very light or white with minimal evidence of a tan tinge to the hair or wool. A few breeds, including Karakul and Gotland, have yet a different selection history. Both of these breeds tend to have dark modifiers that cause *Agouti* patterns to have more extensive black regions than the same allele would produce in most other breeds.

Roger Lundie has been the researcher foremost in finding and describing a host of symmetrical patterns that are related to the *Agouti* locus. His list of patterns constantly grows, so any list is likely to be outdated by the time it is published. His website is a useful resource for readers interested in the details: http://www.sheepcoatcolourgenetics.co.nz/an-introduction/. Lundie's interpretation of the sheep *Agouti* locus finely splits the patterns into many discrete and identifiable patterns, each assumed to be the product of a different allele. This is in contrast to Adalsteinsson's early foundational work that grouped similar patterns together. Both approaches have yielded important insights into this locus and its function in sheep.

The *Agouti* locus patterns of sheep are numerous and can be confusing. The various symmetrical patterns are caused by different alleles. The general relationship holds true that "distinct alleles each cause a unique pattern." The converse can be stated as "unique patterns are each caused by a distinct allele." However, this second concept does not always hold true because some alleles produce a range of final patterns. These pattern expressions can be sufficiently distinct from one another that they lead observers to erroneously conclude that they are caused by different alleles. The most accurate linkage of pattern to allele depends on observations of multiple generations within a single family. With an awareness of the potential range of patterns produced in mind, it is possible to explore the *Agouti* locus of sheep. The underlying assumption used in this book is to base most interpretations solely on phenotypic observations. This approach may well infer several alleles for similar patterns when in fact only a single allele is actually acting.

Lundie has further proposed that the many final patterns produced by alleles at this locus result from specific combinations of a set of basic and repeatably expressed "components." Each component is a specific array of tan areas residing on specific body regions. About 15 distinct components have been proposed. These team up in different combinations to produce the various patterns, each of which is attributed to a distinct allele. The general trajectory is that components form alleles either by standing alone or by uniting in various combinations. The alleles then each produce a pattern. The reverse trajectory can be stated as "each observable pattern is produced by a specific allele that is a combination of various components." The logic behind the two pathways is subtly different. The second pathway that starts with patterns and ends with components is a pathway commonly adopted by many observers. This is a useful approach and is likely to be generally accurate, but once again with the caution that related patterns all caused by a single allele will be assigned multiple alleles.

Some subtle and repeatable differences among certain patterns suggest that a single absolute underlying principle that relates these components back to an allele may not always be possible. For example, the fine details of head and leg patterns tend to be consistent within a single pattern, but some of these details are distinct enough from one pattern to another to suggest that they are not from an identical component. Many of the sheep patterns do have highly repeatable components of tan regions, but for some patterns this repeatability is somewhat weaker. Further research into the molecular basis of these alleles will sort through some of these unsolved mysteries.

The *Agouti* locus of sheep, therefore, has three layers of complexity, with complicated relationships among the three. The first layer is the patterns of symmetrical distribution of pigments that are certainly due to this locus. The second layer is the alleles as they act to produce the patterns. The complexity of this interaction resides in whether or not the patterns and alleles always link up in a one-to-one relationship. The third layer is the components and how, or whether, they combine to form the alleles.

The discussion presented here is based mainly on phenotypic details. This approach requires a clear understanding of potential pitfalls that arise from the linkage of specific alleles and specific patterns. The patterns neatly arrange themselves into groups, each of which contains several phenotypes that are similar to one another. It is likely that each of these similar patterns is due to a distinct and independent allele, but some of these patterns could instead be nothing more than variable expressions of a single allele. This latter situation is certainly true for the few highly variable goat *Agouti* phenotypes that have been documented to be the results of single alleles. While the drawings and descriptions of *Agouti* patterns presented in this chapter follow a useful system, it is one that tends to maximally divide the phenotypic patterns and assign each to a different allele. This leads to a higher number of potential alleles rather than a lower number that may be biologically accurate.

A few details of the sheep patterns are especially worthy of note. One is that many patterns include a distinctive pale patch in the preorbital area in front of the eyes. This occurs as a consistent feature and its presence can help to distinguish between some of the patterns. It is also a very useful signal that the *Agouti* locus is being expressed. This is especially the case on sheep that are otherwise black or nearly black, because these small pale areas can separate out the dark sheep that have *Agouti* locus patterns from the sheep that have *dominant black* at *Extension* and therefore lack any *Agouti* locus expression. A second consistent detail that repeats in several patterns is a pale patch on the upper lip at the front of the muzzle, which is similarly helpful in establishing the expression of the *Agouti* locus pattern in sheep.

The *Agouti* alleles of sheep can be arranged, at least loosely, from the most extensively tan to the most extensively black. Along this trajectory they fall naturally into groups consisting of patterns that are broadly similar. In this chapter the patterns are illustrated as drawings in order to capture some of the details more clearly than can be done with most photos. The background silhouette of a fairly generic sheep is chosen in order to include both hair sheep and wool sheep that fully expresses the entire range of tan shades that are possible, although dark tan is likely lacking for a few of the patterns when they are present on wool sheep. Most of these patterns on wool sheep have somewhat less distinct markings, especially on the portions of the body covered by wool. In addition, the tan areas on wool sheep are nearly all changed to cream or white. That difference in expression of tan between wool sheep and hair sheep should be easily inferred from the drawings.

The drawings have a consistent representation of ink colors:

- Black represents eumelanic areas
- Shades of grey are used to illustrate those regions where white fibers grow into black areas with a final result of a grey shade of color. Grey is also used for those areas that are mixtures of grey, black, and white fibers. This is usually true of wool sheep
- Shades of brown are used for areas that have a mixture of both tan and black hairs, or have hairs with tan tips and a black base. Both of these phenomena lead to a dark final appearance
- Shades of tan reflect the usual intensity of pheomelanin in the various patterns
- White is used for the palest tan areas as well as those that are obviously white. This is especially the case for wool sheep.

The following sections divide the patterns into groups. Each pattern is listed as if it were caused by a separate identifiable allele. The pattern names follow the Lundie system. The names are sometimes tricky, because they are acronyms that refer back to some of the specific tan components that are expressed in the pattern. "Saun" is "side and upper neck." "Swiss" refers to facial markings as well as other details. "Sarb" refers to "side and rear back." "Xerus" derives from an African rodent with a pattern similar to the one in sheep. The other names are breed names, flock names, or geographic names that should be reasonably obvious and understandable for the average reader. Not all patterns are exhaustively described in fine detail, but each highlights the main key features that are used to identify the pattern and separate it from other patterns in the group. The groups that the patterns, and their likely alleles, fall into are outlined in Table 6.1.

6.1 Patterns That Are Mostly Tan or White

Nine patterns are nearly entirely white or tan. All of them have a similar final phenotype. For this group especially, it may be the case that these patterns are the result of a single allele rather than each being caused by a different allele. A similar group of phenotypes in goats can be all related back to a single allele that varies in expression as it moves through the generations. The sheep patterns are illustrated in Figures 6.2 to 6.10.

Table 6.1 The color patterns of sheep that can be related back to the action of the *Agouti* locus are numerous. These patterns arrange themselves naturally into six general groups from most extensively tan to most extensively black. This helps to organize their inherent complexity. Remember that "tan" in wool sheep is often expressed as white.

mostly tan or white	tan with black trim	grey that lightens after birth	mixed tan and black	interplay of tan and black	mostly black
white/tan	*wild*	*grey*	*blue*	*xerus*	*black & tan*
tan sarb	*badgerface*	*finn grey*	*light blue*	*cameroon*	*soay*
welsh mtn	*light*	*finn grey eye*	*sides*	*xerus*	*eye patch & tan*
red	*badgerface*	*patch*	*english*	*swiss xerus*	*cariri*
texel red	*torddu*	*grey sarb*	*blue*	*temuka*	*sweep*
swaledale	*badgerface*	*grey eye patch*	*fawn eye*	*enblu & tan*	*lateral stripes*
red	*badgerface*	*light grey*	*dark blue*	*kaoko swiss*	*ouessant swiss*
portland red	*swiss*	*gotland grey*	*gotland*	*swiss eye*	*rusty midsides*
tan with	*hip spot*	*grey and tan*	*dark*	*patch*	*light and tan*
extremities	*badgerface*	*corsican grey*			*sides and tan*
tan spreading	*barbados*	*herdwick*			*tan eye*
white	*blackbelly*	*pale herdwick*			*swiss markings*
dark tan	*barbados*				*eye patch &*
	saun				*sides*
	corsican				*eye patch swiss*
	eyebrow				*eye patch*
	deccani saun				*paddington*
					burrit
					nonagouti

Figure 6.2 The *white/tan* (A^{Wt}) allele results in a completely tan or white phenotype with no black areas. This is the allele responsible for most European breeds that are white, as well as several others. In those breeds the tan is modified to be pale to the extent that it appears ivory or white.

Figure 6.3 The *tan sarb* (A^{tsb}) pattern is almost entirely white, but retains tan on the back of the neck.

Figure 6.4 The *welsh mountain red* (A^{wmr}) pattern has pale tan on the points and on the back of the neck. The wool staple has a white base so the fleece fades with age. The muzzle is white.

Figure 6.5 The *texel red* (A^{txr}) pattern is a pale fawn color with white on the sides of the brisket and the upper portions of the insides of all four legs, as well as the sides and underside of the jaw and chin. The muzzle is white to grey. The center of the inside of the ears is white with a tan edge. The outsides of the ears are tan. The tan tips of the wool fade rapidly as the fleece grows, so by the time sheep are shorn the wool is white or nearly white. Adults are tan on the upper half of the face and legs, but otherwise white.

Figure 6.6 The *swaledale red* (A^{swr}) pattern is likely a synonym for *ouessant red*. It is difficult to interpret the pattern on Swaledale sheep, because in that breed the background color is nearly black where other sheep tend to have tan. This is likely due to dark modifiers at other loci, and is similar to the situation with the Down breeds. The regions with short hair are nearly black, but the fleece is nearly white. A white muzzle and prominent eye patches are key features of this pattern.

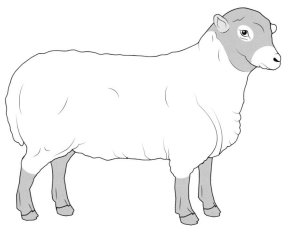

Figure 6.7 The *portland red* (A^{plr}) pattern is typical of Portland and Coburger sheep. The lambs are born a light red color. The muzzle is white and other regions with short hair are pale tan. The fleece rapidly fades to off-white.

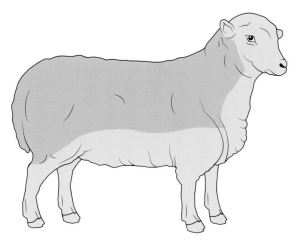

Figure 6.8 The *tan with extremities* (A^{tnx}) pattern occurs in Damara sheep. The overall color is tan, but the color blends to cream or nearly white on the legs and the bottomline of the neck and head, as well as on the lower body.

Figure 6.9 The *tan spreading white* (A^{tsp}) pattern is uniformly tan with white on the inside of the legs, side of brisket and muzzle, around the eyes, and behind the back of the jaw. Homozygotes have more extensive white regions.

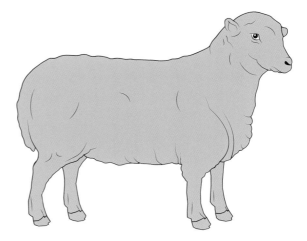

Figure 6.10 The *dark tan* (A^{dtn}) pattern is a darker tan than in *white/tan*, and is seen in the Solognote breed of France and others with dark red face and legs. The fleece is pale and retains a few pigmented fibers. Solognote sheep often shed their fleeces, which then dramatically reveals the red color of the short hair that remains.

6.2 Patterns That Are Mostly Tan with Black Trim or Periphery

Within this group are ten related patterns that all have tan bodies with distinct black patterns on the head, belly, and legs. The tan is often diluted to white on wool sheep. The final phenotype is generally tan with a black periphery in hair sheep, and nearly white with a black periphery in wool sheep. The body, neck, and most of the head are tan or white. The legs, belly, perineum, bottom of tail, underline of neck, and bars on the head are black, as are the bottoms (insides) of the ears in most patterns. Deviations from this general arrangement are discussed under each individual allele. Very few of these patterns are identical to similar patterns in goats, despite some tendency for some authors to list them as the same pattern. These patterns are illustrated in Figures 6.11 to 6.20.

Figure 6.11 The *wild* (*A⁺*) allele causes the color of Mouflon sheep. This is largely tan, which in older rams becomes very mixed with black. The belly and lower legs are nearly white, with nearly all sheep having a black line between the pale belly and the tan body. Some individuals also have thin black stripes down the fronts of the lower legs. The stripes on the front legs break at the knee. The muzzle is pale, as are the insides (bottoms) of the ears, the rump, and the bottom of the tail. There are light tan rings around the eyes and an inconspicuous small bar above the eye. The middle of the side has a pale patch, the shade of which varies with season and age. Lambs are usually a somewhat bland tan color that is nearly uniform, and the black pattern develops with age. The final pattern in rams is much darker and blacker than the pattern in ewes.

Figure 6.12 The *badgerface* (*Aᵇᶠ*) pattern is extensively tan with distinct black markings. The belly is black, extending up the underside of the neck to a black jaw and chin, with some sheep having a break in the black at the throatlatch. The upper lip and central muzzle are black, with some white or tan bordering the nostrils. Black bars on the head begin in front of the eye and go back, above the eye, to meet at the crown and back of the neck at the poll. Black is present on the cheek below the eye, running from the rear of the jaw to the ear. The inside (bottom) of the ears is black, the outside (top) is tan with a black edge. There is a small white bar in front of the eye and the preorbital region is white or tan.

Figure 6.13 The *light badgerface* (*Aˡᵇᶠ*) pattern has more extensive tan regions on the brisket, navel, and scrotum. The body is a light tan or white. The black head markings are well defined and relatively small. The chin, upper lip, and center of the muzzle are off-white.

Figure 6.14 The *torddu badgerface* (A^{tdbf}) pattern has black areas that are crisp and well defined, with tan on the posterior aspect of the lower legs. The bottomline of the neck has a broad black band. The inside (bottom) of the ears is black, and there are small black bars above the eyes. This is a defining pattern for some strains of colored Welsh Mountain sheep, and the crispness of the pattern is partly the result of selection within that breed for a very pale background color accented by the black marks.

Figure 6.15 The *badgerface swiss* (A^{bfs}) pattern is very extensively tan at the expense of any black areas. Only the lower legs are black, with tan on the anterior portions. The black head markings are indistinct and small. The hairs of the belly are black but with tan tips. The black underline on the neck is minimal.

Figure 6.16 The *hip spot badgerface* (A^{hpbf}) pattern has a black belly that extends up the bottomline of the neck, up the perineum to the bottom of the tail, and includes the legs. There is a preorbital white patch. The poll is black. The hip and shoulder are a slightly darker tan than the rest of the body, with some black in these regions. The outside (top) of the ears is black with a tan edge.

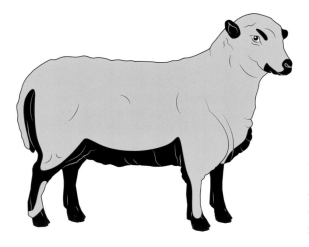

Figure 6.17 The *barbados blackbelly* (A^{brbb}) pattern is extensively tan and the black areas are crisp and distinct. The underline of the neck tends to be tan rather than black.

Figure 6.18 The *barbados saun* (A^{brsn}) pattern has more extensive black areas, but also tan around the eyes and on the cheeks. The middle portion of the abdomen is black, and a black band separates the body from the lateral foreleg.

Figure 6.19 The *corsican eyebrow* (A^{crey}) pattern has less extensive black markings than the others in this group. It has a pale muzzle, and tan or white on the backs of the hind legs and the backs of the front legs. The base of the wool staple is white.

Figure 6.20 The *deccani saun* (A^{dcsn}) pattern has extensive black regions including a black patch on the midportion of the side of the abdomen. The lateral foreleg is tan, and the head is nearly all black with small tan patches above and in front of the eyes.

6.3 Grey Patterns That Lighten After Birth

Eleven patterns can be loosely grouped together as variations on grey. All of them are similar. Sheep with these patterns develop a light staple to the fleece after birth, so they are born dark and become lighter as they acquire the pattern. They also nearly all have a white or nearly white muzzle and patches around the eyes, as well as white rings above the hooves. These pale areas tend to develop after birth but can be present in some lambs (especially homozygotes) at birth. All of these patterns have been proven to reside at the *Agouti* locus by several studies that tracked them through several generations. This is in contrast to the situation in goats where analogous alleles do not occur at the *Agouti* locus. This is one of the important differences in the genetics of the *Agouti* locus of these two species. The final effect of these patterns on a hair sheep background is uncertain, although they do have obvious phenotypes on wool sheep. The phenotypes on wool sheep are illustrated in Figures 6.21 to 6.31.

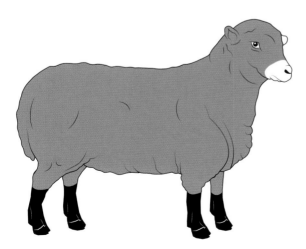

Figure 6.21 The *grey* (A^g) allele in double-coated fleeces results in primary fibers that are black and secondary fibers that are pale. This generally fits with an interpretation of pale pheomelanin. The inside (bottom) of the ears is pale. The muzzle becomes pale, and a pale ring is present above the hooves. There are no light areas around the eyes. These changes are generally present by the age of weaning, and usually start to appear soon after birth.

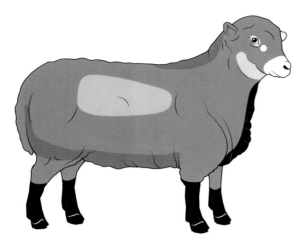

Figure 6.22 The *finn grey* (*A^fng*) has shades of color that are darker on the belly, rear, and back of the neck and along the back, and include a narrow dark band on the underline of the neck. The preorbital area is white. A tan area is present on the rear and sides of the jaw as a bib. The base of the hairs in the dark regions is white.

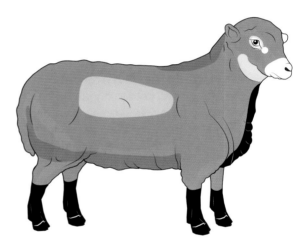

Figure 6.23 The *finn grey eye patch* (*A^fnge*) pattern adds a tan area around the eye to an otherwise *finn grey* coat. The tips of hairs on the body and upper legs are also tan.

Figure 6.24 The *grey sarb* (*A^gsrb*) pattern is similar to *grey* but is mottled on the sides and back. It has a darker area around the neck, and can be darker along the backline, and from the brisket to the naval.

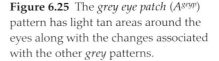

Figure 6.25 The *grey eye patch* (A^{geyp}) pattern has light tan areas around the eyes along with the changes associated with the other *grey* patterns.

Figure 6.26 The *light grey* (A^{lg}) pattern occurs in Gotland sheep, and is also known as *corriedale grey*. These two are likely identical, but occur in different breeds. The Corriedale breed has a lighter final phenotype, the Gotland a darker one which is typical of the general expression of patterns in these two breeds. Lambs are born with obvious tan tips to the hair on the legs. They have a tan eye patch and tan on the outside (top) of the ears and sometimes under the rear of the jaw. Homozygotes are very pale at birth, and both homozygotes and heterozygotes lighten after birth. Corriedales can have a nearly white head.

Figure 6.27 The *gotland grey* (A^{gg}) pattern has a light muzzle, rings above the hooves, white inside (bottom) of the ears, and a light base to the coat. It is somewhat darker over the back and lighter on the belly, bottomline of the neck, navel, and prepuce. There are tan areas around the eye and onto the cheek.

Figure 6.28 The *grey and tan* (*A^gt^*) pattern is an overlap of both *grey* and *black and tan* and has the typical features of both. It is a well-established allele at the *Agouti* locus.

Figure 6.29 The *corsican grey* (*A^crsg^*) pattern is generally lighter than is typical of the other *grey* patterns.

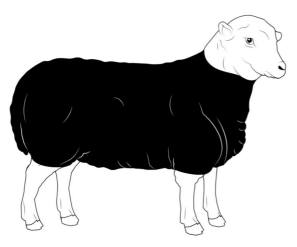

Figure 6.30 The *herdwick* (*A^hrd^*) pattern changes with age and also with season. The birth coat is nearly black. Legs and head become pale as white fibers grow. The body usually remains fairly dark, because the coarse outer coat fibers remain black during the fast growth period in the spring and summer, but become whiter when growth slows in autumn and winter.

Figure 6.31 The *pale herdwick* (A^{phrd}) is similar to *herdwick* but on lambs the body is mottled instead of black. The two Herdwick alleles have the complication of occurring in a breed with a distinct and unique color phenotype that is likely caused by modifiers. These modifiers darken the body coat while causing the legs and head to become increasingly pale.

6.4 Patterns with Mixed Areas of White/Tan and Black

These seven patterns, especially in wool sheep, tend to have both pale and black fibers distributed in different regions on the body. In contrast to grey patterns, these areas are present at birth and do not change very much afterwards. The result in several wool sheep breeds is a fleece that varies in final shade from body region to body region. These patterns are illustrated in Figures 6.32 to 6.38.

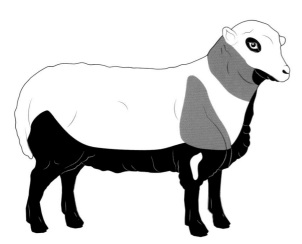

Figure 6.32 The *blue* (A^{bl}) pattern is similar to *badgerface* but with a darker overall color. There is usually a fawn tip on the body and neck fibers. This pattern can have a black band up the bottom of the neck.

Figure 6.33 The *light blue* (A^{lbl}) pattern is whitish-grey on the brisket, navel, and scrotum, and up to the anus to form a small rump patch. The chin is pale, as is the central portion of the muzzle and the inside (bottom) of the ears. Sometimes there is a narrow bar over the front of the eye.

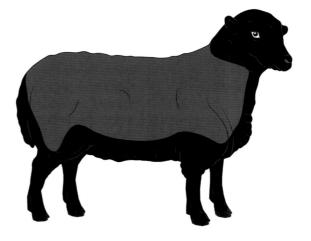

Figure 6.34 The *sides* (A^{sds}) pattern in hair sheep has tan limited to the body, with peripheral areas all black. The head is nearly solid black. In homozygotes there is no black back stripe. The body hair has a tan tip and a black base, leading to the dark color. The *sides* allele is also named D_2A in Soays. It is nearly black with a tan tip to the hairs of the posterior body at birth.

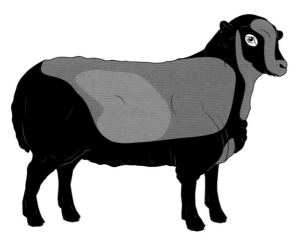

Figure 6.35 The *english blue* (A^{enbl}) pattern is common in Leicester Longwool sheep and other longwool breeds. The pattern is grey on the sides and the rear of the back, and lightest on the upper midsides. The preorbital area is white, and the front of the muzzle is grey. The forehead, across the rear of the nose, and down the back of the jaw are grey.

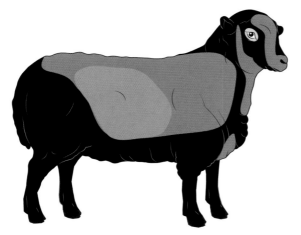

Figure 6.36 The *fawn eye* (A^{fney}) pattern is similar to *english blue* with the addition of tan areas around the eyes and on the cheeks. It often has a black line down the back, and tan tips to body fibers and on the outside (top) of the ears.

Figure 6.37 The *dark blue* (A^{dkbl}) pattern has black on the underside of neck, a white preorbital area, and a grey muzzle area. The jaw is pale extending up to the base of the ears and the central forehead. The body is grey and the backline, limbs, belly, perineum, and tail are black. The fibers on the sides of the body and neck have tan tips.

Figure 6.38 The *gotland dark* (A^{gtdk}) pattern occurs in the Gotland breed and is modified to be dark, resulting in a grey body and black legs and head. There are small white areas on the lower and upper lips, and the center of the front of the muzzle.

6.5 Patterns with an Interplay of Tan and Black Areas

These seven patterns have both tan and black areas over the body. This is similar to the previous group, but usually with more distinctly tan regions. These patterns are different from most others in which the body is generally either tan or black but not a combination of the two. Several of these patterns occur in wool sheep, and the resulting fleeces have various

Figure 6.39 The *xerus* (A^{xrs}) pattern has a tan patch on the middle of the side of the body, and tan areas on the lower legs, bottomline of the neck, bottom of the jaw, inside (bottom) of the ears, and small bars above the eyes. There is a small white patch in the preorbital area. Some breeds with this pattern have lower legs with tan anteriors and tan posteriors surrounding black in between. This is distinctive, and may point to an allele separate from the one named here.

Figure 6.40 The *cameroon xerus* (A^{cmxr}) resembles *xerus* but is more extensively tan. Black is present on the head, and a wide band along the topline, laterally on the shoulder, and spreading down from rump along upper leg to the lateral thigh. Tan is found on the inside (bottom) of the ears, and also as bars above the eyes.

Figure 6.41 The *swiss xerus* (A^{swxr}) is more extensively black than *xerus*. It has tan lower legs, belly, inside ears, bottomline of neck, middle of sides, and sides of thighs and lower rump to connect to the rump/perineal patch. The preorbital area is white, and is usually included in the bars that run from the muzzle to above the eyes.

Figure 6.42 The *temuka* (A^{tem}) pattern is also named N_5A in Soays. Lambs can be born nearly tan but with a black base to the hairs. The body is pale with a darker shoulder and rump, black legs, and a black head with a tan eye patch. Tan is present on the outside of the legs down to hooves.

Figure 6.43 The *enblu and tan* (A^{eblt}) pattern has a complex interplay of areas. Pale areas include the dorsal aspect of the abdomen, belly, underline of the neck, bottom of jaw, rump, and preorbital gland. Black areas are observed on the shoulder and neck down to the knees, on the lower aspect of the sides, and then up to the croup and down to the thighs.

Figure 6.44 The *kaoko swiss* (A^{kksw}) is generally pale or tan, with black areas around the eyes, and backs of legs, and a sooty appearance over the body. In hair sheep the result is a tan sheep with black areas on the neck and shoulder, down to the lateral foreleg. The outsides (tops) of the ears have a black longitudinal bar, and there is black around the eye, on the cheek, and across the bridge of the nose. The belly and inside of the legs are pale.

Figure 6.45 The *swiss eye patch* (*A^swey*) has light areas that include the underline of the neck extending up along the back of the jaw, bars from muzzle to eye, insides of ears, belly, lower sides, and thighs. Darker areas include the shoulder, and along the topline to the croup, and extend down to the stifle. Black persists on the fronts and backs of lower limbs, bridge of nose, and lateral lower portion of the upper foreleg.

shades of color depending on the body region, and have a strong appeal for hand spinners and other craftspeople. On hair sheep these patterns retain a more distinctive distribution of tan and black areas. These patterns are illustrated in Figures 6.39 to 6.45.

6.6 Patterns That Are Mostly Black or Have White or Tan Trim

The most extensively black patterns include 18 different ones, each with its own specific distribution of tan regions. The tan regions tend to be on the head and legs, although a few of these also have tan regions on the body. The final phenotype of most of these is one that produces a black or nearly black fleece. These are illustrated in Figures 6.46 to 6.63.

Figure 6.46 The *black and tan* (*A^t*) pattern is common across several breeds. It has a tan belly, small tan areas above and in front of eyes, inside the ear, and on the rump and the underside of the tail, a small area at the throatlatch and up the center of the chin, and on the front of the upper lip, as well as on the rear aspect of the lower legs. This pattern is common in several breeds worldwide. On wool sheep the tan tends to be white or nearly so.

Figure 6.47 The *soay* (A^{soay}) pattern is similar to *wild* but is darker and more somber due to a black base to the body hairs.

Figure 6.48 The *eye patch and tan* (A^{eypt}) has the tan areas of the *black and tan* pattern with the addition of the tan patch around and below the eye.

Figure 6.49 The *cariri* (A^{car}) pattern has more extensive tan areas than *black and tan*. The markings on the front legs have no break at the knee. It is named after the Cariri sheep breed in Brazil.

Figure 6.50 The *sweep* (A^{swp}) pattern is a common allele in Leicester Longwools and other longwool breeds. It has a white preorbital area, a pale upper lip in the muzzle region, and a pale area on the middle of the sides. The extent of the pale areas can drastically change the overall color of the fleece that is produced by this allele. Many sheep have a pale area on the upper forelegs.

Figure 6.51 The *lateral stripes* (A^{ls}) pattern has a white upper lip and muzzle, and a pale bar from muzzle to eye but which does not extend beyond it. A pale stripe runs between the belly and the side. The underside of the neck is pale, and there is some shading on the body.

Figure 6.52 The *ouessant swiss* (A^{ousw}) pattern occurs in that rare French breed. It has a pale muzzle with a bar up to the preorbital area, the underside of neck, and lower flank, frosting on belly, and pale or tan on the midside of the body.

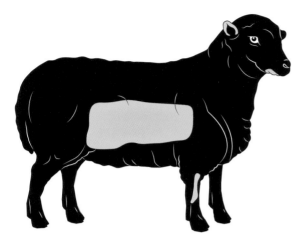

Figure 6.53 The *rusty midsides* (A^{rsmd}) pattern has tan tips on the fibers that are on the midside of body, the rear of the side of the jaw, the outside tips of the ears, and the outside and front of the front legs up to knee height. The preorbital area is white, and many sheep have traces of tan between the ear and eye.

Figure 6.54 The *light and tan* (A^{ltn}) is similar to *black and tan* but with more extensive tan areas.

Figure 6.55 The *sides and tan* (A^{sdtn}) is also named D_5T in Soay sheep. It has tan areas similar to the *black and tan* pattern, but also has tan tips to the hairs on the body.

Figure 6.56 The *tan eye* (A^{tney}) pattern is similar to *lateral stripes* but also has tan around the eyes and on the cheek.

Figure 6.57 The *swiss markings* (A^{sw}) pattern has a frosted belly, an extensive pale area on the underside of the neck, and rear of the flanks, and prominent pale head bars. This pattern is often held up as analogous to the *toggenburg* pattern of goats, but is quite distinct from it. Many of the patterns in sheep that are otherwise superficially similar to patterns in goats show important differences from the goat patterns. Very few of the patterns seen in these two species are identical.

Figure 6.58 The *eye patch and sides* (A^{eysd}) pattern is one of the variants in Soay sheep. It is dark, with tan-tipped sides, and pale eye and cheek patches.

Figure 6.59 The *eye patch swiss* (*A^eysw*) pattern combines two patterns, resulting in a tan patch around the eyes and on the cheeks, and a white moustache area on the front of the upper lip and chin. It also has a frosted belly.

Figure 6.60 The *eye patch* (*A^eyp*) pattern has tan areas around the eyes and on the cheeks below the eyes.

Figure 6.61 The *paddington* (*A^pad*) pattern darkens other patterns when heterozygous with them. It segregates as an allele at the *Agouti* locus. The pale areas include the base of the ears, small dots on forehead, and a small pale area on the throatlatch. The final result is nearly black.

Figure 6.62 The *burrit* (A^{brrt}) pattern has a pale belly, and insides (bottom) of the ears. These areas are similar to *black and tan*, but this pattern lacks the other tan marks of *black and tan*.

Figure 6.63 The *nonagouti* (A^a) allele produces a phenotype that is black or nearly so. A few of these sheep have minor leakage of tan onto the bottom eyelid and the rear half of the top eyelid. This leakage has been most noted in breeds that lack pigmented skin and have white hooves.

A complete list of sheep *Agouti* locus patterns is nearly impossible to assemble, because new patterns are being named frequently. By the time any list is published it is very likely that several patterns will have already been added and will fail to appear on that list. It is challenging to sort out the fine details of the *Agouti* locus in sheep, but a few general tendencies help breeders to use variation at the locus to their advantage. The general rule regarding the codominant expression of pheomelanic areas (tan in hair sheep, pale cream to white in wool sheep) can be very useful. The specific leg and facial patterns can also help to identify patterns in many cases, and these areas can be a more reliable indicator of *Agouti* genotype than is sometimes betrayed by the body.

Most of the *Agouti* locus alleles in sheep have been much more thoroughly studied in wool sheep than in hair sheep. Hair sheep present a few theoretical problems to those interested in the *Agouti* locus. In wool sheep the *Agouti* locus alleles each tend to cause a very definite and reproduceable pattern. To put it another way, nearly all wool sheep having

pale to produce the ideal dark blue face color. Selecting away from the homozygous *white/tan* genotype resulted, at least traditionally, in a very high frequency of *white/tan* sheep in these breeds that were heterozygous for other *Agouti* locus alleles. This selection pressure was likely most notable in the Wensleydale, but was also present in the other breeds.

The consequence of selectively retaining heterozygotes is that these breeds produced a high frequency of colored lambs. This was long true throughout the many years when these breeds were selected to have white wool and bluish faces, even in the face of the culling of colored lambs. As an example, the Wensleydale breed in the early 1900s produced something in the order of 20% colored lambs, which indicates that most Wensleydale sheep in that time period were heterozygous for recessive *Agouti* locus alleles. Over the years, breeders of registered white Wensleydale sheep discovered that if they became a bit more forgiving of the sheep that were "too white" they could reduce the percentage of colored lambs in the breed. The consequence of this change was an increasing number of homozygous *white/tan* sheep participating in the breed. As a result, more recent years have seen a decline in the production of colored lambs compared to that of a century ago. A final shift in the frequency of alleles for color in these breeds is that the registries of many of the longwool breeds now allow inscription of colored sheep. The breed registries vary on exactly how to manage these colored sheep within the breed. Some registries maintain a separate subsection for them, others afford them full inclusion into the breed registry along with the white sheep.

The longwool breeders' insistence on a blue tinge underlying their white sheep had other subtle consequences. Insisting on blue color meant that white sheep with an underlying *brown* homozygous genotype at *Brown* would likely be culled. Over years of selection this has led to a very low frequency of the *brown* allele in these breeds. The combination of extensively black *Agouti* locus patterns, along with rare or non-existent *brown* alleles, has the result that these breeds tend to fairly commonly produce sheep that have black or silvery-grey fleeces of varying shades, but only rarely produce sheep with fleeces of brown shades. Many producers of colored wools want a range of colors due to the demand from hand spinners, and this can include either individual fleeces with variable shades, or single shades of color throughout an individual fleece. The variable colors within a single fleece are routinely produced by *Agouti* alleles such as *blue*, *sweep*, and *english blue* that are common in these breeds (Figure 7.4). In most flocks that express these alleles the final shade of wool varies not only over the body of the individual sheep, but also from sheep to sheep (Figure 7.5). The result is that a flock with these alleles can provide a wide range of fleece colors.

Breeders of longwool sheep that desire dark or nearly black fleeces can find success to be a challenge. In these breeds the *nonagouti* allele does occur, but is fairly rare. Once this allele is located in a sheep, that sheep can be mated with the other colors (all of them usually greyer in tone) to produce heterozygotes. A clue to the presence of the *nonagouti* allele is that heterozygotes are often a bit darker (on average) than the homozygotes for other alleles and those that lack *nonagouti* (Figure 7.6). In addition, sheep expressing a more extensively black pattern are more likely to carry *nonagouti* than those expressing a less extensively black pattern. The "most black" patterns can only be homozygous or mask *nonagouti*, while the "least black" patterns can be homozygous or can mask multiple other alleles in addition to *nonagouti*.

Figure 7.4 Most colored Leicester Longwools have variable color over their bodies.

When heterozygotes are mated together they produce the desired *nonagouti* black homozygous lambs at a rate of about 25%. This is outlined in Table 7.1. The strategy of mating heterozygotes together works well, but needs to be fine-tuned for long-term success if the goal is the consistent production of *nonagouti* homozygotes. Mating homozygotes together usually involves some degree of inbreeding, because in most cases these will have been produced by the same few parental sheep with this rare allele. It is possible to avoid inbreeding by assuring that at least some of the *nonagouti* homozygotes are mated to unrelated *blue* or *sweep* mates to produce heterozygotes that are less closely related to the previous generations. These heterozygotes can then be used in the next generation to produce black lambs with a lower level of inbreeding.

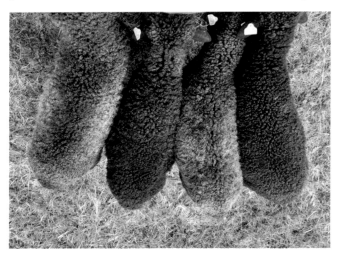

Figure 7.5 Individual Leicester Longwools vary in their final color, even with similar genotypes. These sheep are all *wild* (black) at the *Brown* locus, and the brownish tint is due to weathering as it is only present at the tips of the fleece (photo by D. Kauffman).

Figure 7.6 These two Leicester Longwool ewes have a similar fleece color. The ewe in A has *blue*, the ewe in B has *sweep*. The *sweep* allele's head pattern is darker, making it less able to mask other alleles. Consequently this sheep is likely to be either homozygous *sweep*, or heterozygous *sweep* and *nonagouti*. In contrast, the lighter pattern on *blue* can mask several other alleles and is therefore a less logical choice if the goal is to maximize the production of dark fleeces.

Table 7.1 Results of using dark patterned longwool sheep to produce black lambs.

	color	*Agouti* genotype	% of lamb crop
sire	dark shaded	"pattern"/ *nonagouti*	
dam	dark shaded	"pattern"/ *nonagouti*	
lambs	lighter shaded	"pattern"/"pattern"	25%
	dark shaded	"pattern"/ *nonagouti*	50%
	black	*nonagouti* / *nonagouti*	25%

Breeders desiring dark moorit colored wool in a longwool sheep breed have additional challenges. Moorit lambs only rarely occur in most of these breeds. Moorit-based lambs have rarely appeared in the Lincoln and Wensleydale breeds, and may well occur rarely in some of the other breeds of the longwool group. Nancy Irlbeck has been instrumental in bringing this color to the fore in the Wensleydale breed. Moorit lambs of these breeds usually have the *sweep, blue,* or *english blue* pattern, and the result is a fairly pale shaded brown rather than a dark, deep moorit color. The pallor diminishes the distinctiveness of the final fleece color. These alleles, when on a black base color, provide for lustrous shades of silver and grey. On a brown background they are paler and, while still lustrous, they lack some of the distinctiveness of the silver shades.

In order to generate dark moorit fleeces the truly black sheep of these breeds can be very useful. The moorit-based patterned sheep can be mated to a black mate, as outlined in Table 7.2. The result is lambs that are heterozygous for *nonagouti* and another pattern, and also heterozygous for *wild* (B^+) and *brown* (B^b) at *Brown*. These sheep are black-based, and also have an *Agouti* pattern with shading across the body. When mated to one another, the expectation is that about 25% of the lambs should be moorit-based, and 25% of those will be the dark moorit with no pattern. This has a somewhat disappointing final yield of only 6.25% of the lamb crop being dark moorit, so obviously the only way to assure success is

Table 7.2 Producing dark moorits in longwool breeds is often a multigenerational process. The table outlines the results from an initial cross of a light and shaded moorit ram and black ewes. Mating the original sire to his own daughters provides the highest yield of moorit-based lambs (light or dark). Mating the black sons and daughters of the initial cross yields the highest proportion of dark moorit lambs. Other matings are possible between the various colors of lambs in the initial cross, but they each reduce the final yield of dark moorit lambs.

	color	*Agouti* genotype	*Brown* genotype	% of lambs
sire	shaded moorit	"pattern"/"pattern"	*brown / brown*	
dams	black	*nonagouti / nonagouti*	*wild / wild*	
lambs	shaded grey	"pattern"/ *nonagouti*	*brown / wild*	100%
sire x black daughters				
lambs	shaded grey	"pattern"/ *nonagouti*	*brown / wild*	50%
	shaded moorit	"pattern"/ *nonagouti*	*brown / brown*	50%
black sons x black daughters				
lambs	black	*nonagouti / nonagouti*	*wild / wild*	25%
	black	*nonagouti / nonagouti*	*brown / wild*	50%
	moorit	*nonagouti / nonagouti*	*brown / brown*	25%

to raise lots of lambs! Once the desired dark moorit lambs are produced they can be used among themselves to produce more moorits, and can also be used with black mates to generate heterozygotes, and dark ones at that. These can then be used to produce yet more dark moorits.

The ideal mates for the moorit sheep in this situation are black sheep, or at least very dark ones. This strategy is better than using a mate with a more dominant, and therefore paler, *Agouti* locus pattern. The challenge is to eventually combine the recessive *nonagouti* allele, that lacks any light pattern, with the recessive *brown* to produce the desired dark moorit sheep. Once a dark moorit sheep is obtained it can be mated with sheep of just about any of the *Agouti* patterns to produce lambs with a full range of black and grey fleece colors. Each of those lambs is a carrier of *brown*, and is sure to produce the moorit-based colors if paired correctly with sheep that also carry the *brown* allele in either the heterozygous or homozygous condition.

In breeds where moorits are rare they tend to occur in only a few families. The result is that nearly all mating strategies that use moorits with one another involve some level of inbreeding. This is not necessarily all that risky for an initial generation, but over several generations it can lead to significant declines in vitality. To avoid long-term inbreeding it is possible to use the moorits among themselves (inbreeding), but also mate them with black or patterned sheep that are unrelated (outbreeding). The moorit to black matings produce heterozygotes with no inbreeding. Through this strategy the frequency of the *brown* allele can be increased throughout an expanding pool of sheep that are only distantly related. This provides for a low inbreeding coefficient when they are mated to one another. Mating heterozygous carriers of *brown* can always bring the moorit color back to expression in the phenotype. The key to the strategy is to be sure that heterozygotes are available, and that these have minimal relatedness.

Figure 7.7 This Shetland sheep has a very dark moorit color that comes from being homozygous *nonagouti*, and homozygous *brown*, as well as having modifiers that intensify the color and make it resistant to fading. The effect of those modifiers on a black background color can be seen in the sheep in the background.

7.1.5 General Strategies for Moorit and Mioget

The *nonagouti* allele occurs in many breeds. It has the potential to provide uniformly black wool, although in most breeds the black color produced by the homozygous *nonagouti* genotype tends to fade dramatically. The result is that the shorn wool is either grey or a shade of brown. Moorit (truly brown) wool follows this same tendency to fade, but tends to fade even more readily than the black wool of an otherwise similar genetic background. In many breeds it is quite challenging to produce dark moorit wool that does not fade (Figure 7.7).

A helpful step along the pathway to moorit sheep is to locate and document the moorit genotype. This genotype can be hidden in white sheep and it is difficult to know whether their base color is black or moorit under their white fleeces. Knowing that a white sheep is a *brown/brown* homozygote can help a breeding program along because these sheep provide increased genetic diversity, while not contributing any *wild* alleles at the *Brown* locus. One strategy.that has worked for some breeders is to examine the eyes of sheep, looking for the subtly lighter color typical of *brown* homozygotes. Another useful hint can be skin color. White sheep over a moorit base color do not have black skin, but of course this detail is of limited help in the many white breeds that have pink skin. In longwools, and other breeds with dark skin, the difference between the black skin of black-based sheep and the brown skin of moorit-based sheep can be a useful indication of the underlying genotype. In addition, some breeders have had success with the "flashlight test" (or "torch test"). This involves shining a light beam in the faces of a flock at night, and noting which sheep have eyes that reflect back a red color instead of the usual green color. This test is not 100% accurate, but can help to locate moorit-based white sheep in many breeds.

White sheep that are based on moorit are useful because they can be mated to phenotypic moorits. The result in most cases is white lambs, but these all carry a recessive *Agouti* locus allele under the white, inherited from the moorit parent, and are also *brown* homozygotes. These white lambs can then be mated to moorits for the production of a lamb crop that produces 50% moorit lambs. The other 50% are white but they all are homozygous for *brown*, as well as heterozygous for the more recessive *Agouti* allele. They therefore have great potential for color production under their white coat if they are mated to other moorits.

The strategy of identifying moorit-based white sheep helps to increase the numbers and diversity of moorit sheep. However, using these white sheep in a breeding program does little to assure relatively dark manifestations of moorit color, because little is known about the underlying *Agouti* locus alleles, or any other modifiers that affect the final fleece color. If the goal is nonfading dark moorits, then one very useful strategy involves mating moorit sheep to nonfading black sheep. This strengthens and darkens the moorit color of any *brown* sheep that are produced. Keeping nonfading black sheep can be counterproductive in a few situations. One example is when truly black fleeces are less desirable than shaded grey ones due to the demand from hand crafters. Unfortunately, if dark moorit is a goal, then nonfading black sheep are a nearly essential ingredient of the breeding program. If the final goal is to produce dark moorits, but also to have somewhat greyer fleeces on the black-based colors, then these two goals pull in opposite directions and cannot be fully met within a single population that does not have the other, less desirable options including jet black and shaded light moorits.

Mating nonfading black sheep to any "too light" moorit sheep produces black-based lambs that are heterozygous for *brown*, heterozygous for *nonagouti*, and also have which-ever modifiers help to keep the wool dark. One way to bring out the recessive brown color after an outcross requires inbreeding back to the moorit parent for a 50% yield of moorit-based lambs. An alternative is to inbreed within the heterozygous half-sibling, black-based lambs for a 25% yield of moorit-based lambs. The second strategy has a lower yield of moorit lambs, but an increased chance that they have all the modifiers that will keep the color dark. Final success may take another generation or two of carefully monitored mating. The lambs produced by this strategy are sure to pick up the *brown* allele, as well as modifiers that lead to less fading. As long as the mates have minimal evidence of fading, the appropriate modifiers are likely present and make it less likely that fading will occur.

In breeds where fading is minimal (some Shetlands are one example) the mioget modification can be important in yielding a range of fleece colors (Figure 7.8). This modification, when combined with both black and moorit, can yield a range of fleeces that run from black, through charcoal and pewter, dark brown, moorit, and the honey color of mioget. The mioget color has other advantages over the usual faded moorit, because it has a yellower tinge and is more of a warm honey color than a cooler light brown.

7.1.6 Moorit Romeldale Sheep

Many modern breeds are the result of deliberately combining finewool breeds with long-wool breeds. These include the Corriedale, Romeldale, Columbia, and others. These breeds often have a great deal of variation in *Agouti* and other color alleles because they have combined the array of alleles from all of their founding influences. The recessive alleles for color in these breeds give breeders interested in colored wool a broad range of options to work with. Romeldale breeders, especially, have embraced this variation. They use it creatively to generate a wide range of colors in fleeces that meet high demand from hand spinners due to the breed's fleece character that includes softness as well as ease of hand processing (Figure 7.9).

Figure 7.8 The interactions of black, mioget, and moorit can provide a wide range of final fleece colors as shown in the Shetland sheep in the photo (photo by L. Wendleboe).

The Romeldale breed was developed from Rambouillet and Romney parents in the early 1900s. Romeldales retain the color alleles and modifiers from both of those breeds, and as a result have a very wide range of colors that many breeders now target. Over the years several *Agouti* patterns have surfaced. The majority of colored sheep that have arisen in the breed are *badgerface* at *Agouti*, with a few other *Agouti* alleles also in evidence. Romeldale sheep most often have the *wild* allele at *Brown*. The result of these two loci in this breed is a predominance of light grey fleeces in the colored sheep. An occasional light or shaded mioget sheep also occurs in the breed, and these are B^bB^b, $Mod^{mod}Mod^{mod}$. They generally

Figure 7.9 The genetic background behind the Romeldale produces a wide variety of colors in the breed (photo by M. Minnich).

Figure 7.10 Many of the moorit-based sheep that pop up in the Romeldale breed are miogets on a *badgerface* background and are therefore quite pale. The contrast between black-based and mioget-based *badgerface* sheep can be seen in this pair (photo by M. Minnich).

also have one of the more dominant *Agouti* alleles with the result being a fairly pale fleece (Figure 7.10). Dark moorits only rarely pop out spontaneously in this breed, so producing them takes a multistep procedure that can be guided by a basic knowledge of the various factors available.

The breeder of one Romeldale flock, Marie Minnich, noticed a *black and tan* ram lamb with good qualities besides his dark color (Figure 7.11). This color results from one of the more recessive *Agouti* locus patterns, and it is logical to conclude that the ram is either A^tA^t or A^tA^a. These two alleles are equally useful for producing moorit, because sheep with either of these alleles will produce black fleeces. A careful look at the other sheep in the flock can be useful in determining which of these genotypes is more likely. If few sheep have the tell-tale light eyebrow patch typical of A^t, then it is most likely that the ram lamb is heterozygous for A^a. This ram is a very clear black, with no hint of brown, so he is most likely B^+B^+ Mod^+Mod^+. Also within the flock are occasional mioget animals, most of them with the genotype of $A^{bf}A^{bf}$, B^bB^b, and $Mod^{mod}Mod^{mod}$. This combination gives a very pale final color. These genotypes are outlined in Table 7.3.

Table 7.3 Initial available genotypes in a Romeldale flock. A^{bf} is *badgerface*, A^t is *black and tan*, A^a is *nonagouti*, B^+ is *wild*, B^b is *brown*, Mod^+ is *wild*, and Mod^{mod} is *modified*.

	color	*Agouti*	*Brown*	*Modified*
ram	black and tan	A^tA^a	B^+B^+	Mod^+Mod^+
ewes	light mioget	$A^{bf}A^{bf}$	B^bB^b	$Mod^{mod}Mod^{mod}$

Figure 7.11 This Romeldale ram has *black and tan* as betrayed by his tan eyebrows. Some of the other markings are from the white spotting that is common in this breed (photo by M. Minnich).

The lambs produced by this cross are all likely to have a light grey *badgerface* pattern. Darker lambs are possible, depending on the presence of any of the more recessive (more extensively black) *Agouti* locus alleles in the ewes. Those will be especially useful if they occur because they will boost the final yield of darker moorit lambs considerably. However, even in the absence of those, the project can proceed and succeed. The lambs can be inspected closely, and by paying attention to eyebrow spots and belly color, the *badgerface / black and tan* lambs (light bellies) can be distinguished from any *badgerface / nonagouti* lambs (black bellies). One of the ram lambs, of either genotype, can be mated back to the original light mioget ewes. Selecting one that carries *black and tan* has an advantage because the allele can be tracked in heterozygotes due to the light belly and the eyebrow spots. The *nonagouti* has a disadvantage because it does not leave any tell-tale traces, even though it is a more attractive target in the final color. This initial generation results in a highly variable lamb crop as detailed in Table 7.4.

This strategy produces a mixture of lamb colors. All are *badgerface*, and therefore have fairly light fleeces. Half of the lambs are black-based and not all that useful for this project with the possible exception that they are all obligate carriers of *brown*. They therefore might be useful because they boost the number of carriers available for future matings. About 25% of the lambs are mioget, and 25% are moorit. Among these are the ones with light bellies that are heterozygous for *black and tan*. These can eventually produce the desired dark brown fleece color.

A next step is to tease out the *black and tan* in combination with the *brown* to provide for

Table 7.4 The results of mating a known heterozygous ram back to mioget-based *badgerface* ewes.

color	Agouti	Brown	Modified	% of lambs
sire (grey *badgerface* with light belly)	$A^{bf}A^t$	B^+B^b	Mod^+Mod^{mod}	
dams (mioget *badgerface*)	$A^{bf}A^{bf}$	B^b/B^b	$Mod^{mod}Mod^{mod}$	
lambs				
darker black *badgerface*	$A^{bf}A^{bf}$	B^+B^b	Mod^+Mod^{mod}	12.5%
lighter black *badgerface*	$A^{bf}A^{bf}$	B^+B^b	$Mod^{mod}Mod^{mod}$	12.5%
moorit *badgerface*	$A^{bf}A^{bf}$	B^bB^b	Mod^+Mod^{mod}	12.5%
mioget *badgerface*	$A^{bf}A^{bf}$	B^bB^b	$Mod^{mod}Mod^{mod}$	12.5%
black *badgerface*/ light belly	$A^{bf}A^t$	B^+B^b	Mod^+Mod^{mod}	12.5%
light *badgerface*/ light belly	$A^{bf}A^t$	B^+B^b	$Mod^{mod}Mod^{mod}$	12.5%
moorit *badgerface*/ light belly	$A^{bf}A^t$	B^bB^b	Mod^+Mod^{mod}	12.5%
mioget *badgerface*/ light belly	$A^{bf}A^t$	B^bB^b	$Mod^{mod}Mod^{mod}$	12.5%

dark moorit fleeces, ideally with an absence of *modified* to avoid lighter colors. This can be achieved in different ways, although most of these strategies have a very low yield of the desired dark moorit color. Mating the moorit sheep that are *badgerface* with light bellies to one another produces lambs that are moorit *badgerface* with dark bellies, moorit *badgerface* with light bellies, and moorit with the *black and tan* pattern. These results are outlined in Table 7.5. Any mating among sheep that carry *black and tan* and *brown* will yield at least a few dark moorits, but stacking the odds in favor of either *black and tan* or *brown* by using homozygotes raises those odds. Because the goal is to combine two recessive genotypes, the yield is always relatively low, and therefore the only reasonable assurance of success is to produce a great number of lambs. Once those dark lambs come along though, that doubly recessive genotype is available and can be used to rapidly raise the frequency of both the A^t and the B^b alleles in the flock. By that mechanism, the breeder also raises the percentage of lambs with the desired dark moorit color (Figure 7.12).

Figure 7.12 This dark moorit Romeldale ewe is close to the target color (photo by M. Minnich).

Table 7.5 Results of mating moorit *badgerface/black and tan* sheep to yield dark moorit lambs.

color	Agouti	Brown	Modified	% of lambs
moorit *badgerface* heterozygote lambs	$A^{bf}A^t$	B^bB^b	Mod^+Mod^{mod}	
moorit *badgerface*	$A^{bf}A^{bf}$	B^bB^b	Mod^+Mod^+	6.25%
moorit *badgerface*	$A^{bf}A^{bf}$	B^bB^b	$Mod^{mod}Mod^+$	12.5%
mioget *badgerface*	$A^{bf}A^{bf}$	B^bB^b	$Mod^{mod}Mod^{mod}$	6.25%
moorit *badgerface* / light belly	$A^{bf}A^t$	B^bB^b	Mod^+Mod^+	12.5%
moorit *badgerface* / light belly	$A^{bf}A^t$	B^bB^b	$Mod^{mod}Mod^+$	25%
mioget *badgerface* / light belly	$A^{bf}A^t$	B^bB^b	$Mod^{mod}Mod^{mod}$	12.5%
moorit *black and tan*	A^tA^t	B^bB^b	Mod^+Mod^+	6.25%
moorit *black and tan*	A^tA^t	B^bB^b	$Mod^{mod}Mod^+$	12.5%
mioget *black and tan*	A^tA^t	B^bB^b	$Mod^{mod}Mod^{mod}$	6.25%

7.1.7 General Quality of Colored Wool Sheep

While this book focuses on color, all sheep breeders fully realize that an individual sheep is much more than its color. The evaluation of sheep for potential retention into a breeding flock needs to take into account several factors including color, details of conformation, and productive potential. The quirks of some of the color alleles come into play here.

The *white/tan* allele at the *Agouti* locus is of primary importance in most European breeds of sheep. This allele has the distinction of being the top dominant allele at its locus. The allele in this position in several species has consequences beyond color. It also influences the general development of animals that bear it. The usual consequence is a slight decrease in reproductive efficiency, although this can be countered by good selection for reproductive performance. The *white/tan* allele also tends to produce a slightly greater body size. White sheep in a multicolored flock that are genetically similar, aside from color genes, therefore have an inherent advantage in terms of size and overall scale simply because of the color allele they have. Body size and scale are often the target of selection in sheep flocks, which leads to an increased retention of white sheep as opposed to colored sheep if these production characteristics are the only consideration used for culling or keeping sheep.

The fact that white sheep are typically a bit larger than colored flock mates can be used to good advantage in practical farm situations, even those that target the production of colored sheep. The breeders of most breeds of sheep state that it is unusual for a colored sheep to fully rise up to the standard of a really good white one. This reflects the additional effects the gene for whiteness has on the whole development of the sheep. It is therefore significant that from time to time in most breeds a colored sheep is produced that is indeed up to the standard of the white sheep, even though it is very rare for a colored one to surpass them. Colored sheep that are up to the standard of a somewhat average white sheep are overcoming the inherent disadvantage delivered to them by their color alleles. The significance of these colored sheep that stand up well to the average standard of white sheep is that in all likelihood they are truly superior in terms of overall genetic potential. These exceptional colored sheep can therefore make important contributions to advancing overall quality in a flock. Exceptional colored sheep have good potential for improving not only the quality of colored sheep, but also that of white sheep if they are mated appropriately.

Similar comparative effects are present for moorit-based versus black-based sheep. Moorit-based sheep are generally smaller framed and weigh less than black-based sheep. In keeping with the details concerning the comparisons of white sheep and colored sheep, a similar comparison of black-based versus moorit-based sheep is possible. Any moorit sheep that rises to the general level of quality of black-based flock mates is likely to be truly superior in overall quality and genetic potential. Such a sheep has a great deal to offer, not only to its moorit offspring, but also to its black-based and white offspring.

Some of this plays out with practical consequences in the feral Soay sheep breed. Their North Atlantic island home is a challenging environment for sheep. Soay sheep have a moderate number of different *Agouti* locus alleles. In general the *nonagouti* homozygotes survive less well than other genotypes. Similarly, the *brown* homozygotes are smaller and less viable than the black-based colors in those years with good forage production. However, in years in which low forage production or other environmental challenges put increased stress on the sheep, these trends reverse. In such lean years, the genotypes that are usually at a disadvantage have a considerable survival advantage. Those years come around frequently enough that all of the various genotypes survive in the sheep flock, with different genotypes favored in different years. This is all due to the influences of these color alleles on traits other than color.

7.2 Strategies for Color Production in Hair Sheep

Strategies for hair sheep depend on the breed involved. Some hair sheep breeds, such as the St. Croix and Barbados Blackbelly, traditionally have only a single color pattern. This restriction is changing somewhat in some breeds, with the result that color now varies more than it did in the past. Some breeds of hair sheep have always been allowed to vary widely in color. In the historically variable breeds, such as the Damara of South Africa, some breeders have the goal of producing dramatically colored sheep for the eventual use of their hides as pelts.

Color in hair sheep is often determined by the *Agouti* locus alleles. If the various patterns are desired, and especially if a wide variety of these are desired, then it can be helpful to mate the sheep that have the more tan alleles (more dominant) to either *nonagouti* or another extensively black (more recessive) pattern. This provides for a high probability that any recessives hidden in tan sheep can be brought to at least a heterozygous condition and can therefore be identified and used if so desired. Through this strategy it is possible to tease out any alleles that might have been long hidden by virtue of being heterozygous with more dominant alleles. Any sheep with *dominant black* can frustrate this strategy. If desired those sheep can be avoided by insisting that the black sheep used for breeding have two non-black parents. They are then assured of lacking the *dominant black* allele.

Spotting patterns can also be important in hair sheep. Dramatic and bold patterns are usually the ones most desired (Figure 7.13). It is possible to combine various spotting patterns together on a single sheep. Usually these combinations produce a sheep that is predominantly white and that therefore might not have a very striking pattern because little background color is present to provide contrast (Figure 7.14). The bold character of the patterns can be missing even though the causative alleles are very much present. To avoid

CHAPTER 8

Cattle Color

Cattle generally have a short hair coat made of primary fibers, and therefore many of the details that are important in colors of fleeced goats and sheep do not affect cattle. The genetics of cattle color has been greatly helped by recent successes in mapping the entire bovine genome. This has provided answers to the exact chromosomal site of many loci that affect color in cattle, and details about the specific site often reveal the precise biological mechanism behind many of the colors.

The registries for most standardized cattle breeds have historically been very restrictive in their allowance of color variation and many breeds had only a single color combination allowed. This still holds true in many breeds, although for several other breeds a recent trend goes in the opposite direction by allowing colors that were historically not typical of the breed. Several of the breeds that were originally red-based now allow black cattle to be registered after a process of upgrading. This has resulted in many breeds that were once red now becoming nearly entirely black. Registries of a few landrace breeds, including Texas Longhorn, Pineywoods, Florida Cracker, and Nguni cattle, stand in contrast to the general trend of a registry permitting only a narrow range of color (Figure 8.1). Registries for these

Figure 8.1 Florida Cracker cattle are a landrace that exhibits much color variation.

Table 8.1 Major alleles and loci contributing to cattle color.

Dominant Red	Extension locus	Agouti locus	dilution
dominant red (rare) wild	dominant black (common) black/red (rare) wild (rare in most breeds) red (common) red charlie (rare)	brindle other patterns nonagouti (black) (rare)	brown charolais dilution simmental dilution tan dilution chinchilla Larson blue albino chediak higashi

breeds allow a wide array of colors. It is in these breeds that color variation is more easily noticed and appreciated.

The genetic control of cattle color follows the same general rules as other species, although loci that are relatively unimportant in most sheep and goat breeds are very important in cattle. The opposite is also true, so that loci that are extremely important in understanding the color of goats and sheep play only minor roles in cattle. The *Extension* locus in cattle is very important, and controls most of the expression of entirely black and entirely red colors. In contrast, the *Agouti* locus of cattle has minimal importance in most cattle breeds. Several different genetic mechanisms behind dilution are also important in cattle, and are better documented than similar mechanisms are in other species. Cattle color is controlled by several loci along with a few other general modifications of base colors. These loci are outlined in Table 8.1, and each is detailed in the following sections.

8.1 Extension *Locus*

Extension locus alleles are the most important determinants of basic coat color in many breeds of cattle. The most dominant allele of this locus is *dominant black* (E^D) which results in a completely black coat. This is by far the most common cause of black cattle worldwide, and any exceptions to this rule are noteworthy due to their rarity (Figure 8.2). Black cattle are common in many breeds worldwide, most notably Angus and Holstein. The occasional red calf is still produced in most breeds that are usually black. Some cattle are dark brown or "off black" instead of the expected jet black caused by *dominant black,* and many of these are heterozygous for *dominant black* rather than homozygous. Most of these are still black enough to be classed as black even by relatively stringent criteria.

Some breeds and breed crosses have genetic modifications that change the *dominant black* phenotype to a very definite brown or even red. This most commonly occurs in the hybrids produced by mating some of the south Asian zebu breeds with European breeds. The generations produced from intermating hybrids from these breeds can vary widely in final color, even though the animals are all heterozygous for E^D. Some of these could barely be classified as black.

The most recessive *Extension* allele in cattle is *red*. This allele is very common, and is analogous to the allele that causes the chestnut color in horses. These cattle are entirely red with a red or white switch to the tail (Figure 8.3). They have no black in them, which is an

Figure 8.2 Black cattle of most breeds have the *dominant black* allele at *Extension,* as evident in this Randall Lineback steer.

important detail. This is a common color in cattle, and accounts for the red of Shorthorns, Red Angus, Red Poll, Devon, and several other breeds.

Simmental cattle that are red usually have the *red* allele, but a separate recessive allele, *red charlie,* also occurs. The *red charlie* allele is due to a different mutation, but the final red phenotype is identical to *red.* The two mutations have a similar mode of action and achieve their identical phenotypes through the inactivation of the receptor produced by the *Extension* locus.

Cattle of many breeds are black, those of several other breeds are red. In most cases it is the two specific alleles, *dominant black* and *red,* that have been used to accomplish

Figure 8.3 Red cattle, like this Red Devon cow, are due to *red* at *Extension* and nearly always have a white tail switch (photo by J. Beranger).

the consistent color in these breeds. It is noteworthy that *red* in cattle is a common allele, because this allele is rare in both sheep and goats, to the point that any occurrence in those species is a novelty.

The cattle *Extension* locus has two more alleles in addition to the two that have so commonly been used to establish uniform color in many breeds. The *black/red* (E^{BR}) allele is recessive to *dominant black* and dominant to *red*. Cattle that express this allele are born a uniform red, but darken to uniform black by about six months of age. The progressive change of hair color could easily result in misclassification if the observation is only made at a single point in time.

The *wild* allele (E^+) is a neutral allele that allows expression of the *Agouti* locus. It is dominant to *red* and recessive to the other two alleles. Most of the *Agouti* locus patterns that are allowed to be expressed by E^+ have obvious black and tan areas, so they can be easily distinguished from the phenotype produced by the other *Extension* locus alleles that cause uniform colors. Colors based on *Agouti* variation occur breed-wide in many breeds such as Jersey, Brown Swiss, and several zebu breeds. All of those breeds have the *wild* allele at *Extension*.

8.2 Dominant Red *Locus*

Holstein cattle have recently experienced a novel dominant mutation that causes a red phenotype which masks all alleles at the *Extension* locus, including *dominant black* (E^D). This allele resides at the locus named *coatomer protein complex, subunit alpha* (*COPA*). In common parlance the locus is now called *Dominant Red* and contains only two alleles. The mutant DR^{DR} allele first occurred in a Canadian Holstein heifer calf in 1980. It is a recent mutation that is currently limited to Holstein cattle only. The *dominant red* allele causes some inclusion of eumelanin along with pheomelanin which results in a dark shade of red.

8.3 Agouti *Locus*

Agouti locus patterns in cattle are combinations of tan and black areas that are symmetrically arranged over the body, as is true of other species. The patterns in cattle differ from those in goats and sheep in two regards. Firstly the cattle patterns generally lack crisp definition between black areas and tan areas, and secondly they vary more in the extent of black, which progresses in a very orderly fashion from least extensive to most extensive. Along this trajectory there are no clean breaks between clearly identifiable patterns, so cattle can be arranged from most extensively black through most extensively tan with no obvious boundary between patterns at any point. Cattle *Agouti* patterns are therefore difficult to categorize into specific, well-defined patterns with clear and repeatable differences that distinguish them one from another.

The continuity of expression of the *Agouti* locus in cattle raises real questions about the validity of linkage of specific alleles with specific patterns. The entire range of expression goes from nearly all tan to nearly all black. It is likely that several alleles influence this. However, the absence of boundaries between patterns makes the assignment of specific patterns to specific alleles dubious at best.

Figure 8.4 These Criollo Yacumeño bulls illustrate both the tan and black extremes of *Agouti* locus expression, as well as variation in the shade of tan.

The shades of the tan areas in *Agouti* patterns range from very pale cream to very dark red. In most cattle with *Agouti* expression the pheomelanin is an obvious tan and not a deep red, although there are numerous exceptions to this general rule. The pheomelanic areas will be referred to as "tan" in this discussion despite the fact that the final color varies from animal to animal.

The most extensively tan *Agouti* pattern is wholly tan with a black tail switch (Figure 8.4). Depending on the level of tan, some cattle with this pattern can be red and could be confused with recessive red cattle from the *Extension* locus. The tail switch on *Agouti* patterns is usually black which reliably distinguishes them from red cattle with the *Extension* locus mechanism.

Cattle with more extensive black areas retain the black switch, and also develop black regions on the head, and commonly on the fronts of the lower limbs (Figure 8.5). The black

Figure 8.5 This Criollo Saavadreño cow shows the progression of black on *Agouti* patterns that affects the legs and head.

Figure 8.6 American Brahman cattle nearly always have some level of *Agouti* pattern. The symmetry of the pattern is striking, as is the character of the black markings on the lower legs.

areas then encroach onto the neck, shoulder, belly, rear flank, and stifle regions (Figures 8.6 and 8.7). Extensively black animals are almost all black with a lighter tan line down the back, as well as a tan muzzle (Figure 8.4). Very rarely is an animal entirely black from an *Agouti* locus allele. This phenotype can be easily confused with the *dominant black* allele of the *Extension* locus. Such recessive black animals have so far been documented only in the Icelandic cattle breed, but no doubt occur in other breeds also.

Expression of the *Agouti* locus patterns is affected by the sex of the animal. Calves of both sexes with these patterns are born entirely tan and increasingly develop black areas

Figure 8.7 This Criollo Patagónico bull shows one of the more extensively black *Agouti* patterns.

as they age. In the aurochs (the extinct wild ancestor of cattle) the pattern in females likely remained tan, with a black tail switch and minor black on head and lower legs. Bulls, however, became progressively black with age until they were nearly all black but with a lighter tan line down the back, tan around the muzzle, and perhaps tan around the eyes and inside the ears. This extensively black pattern has been used as the pattern of the recreated aurochs bred by zoos, but in those cases it is present in both bulls and cows. The reconstructed aurochs therefore differ from the original wild-type color, which was more tan in females and more black in males.

Most breeds of cattle with the *Agouti* locus expression share the trend that bulls are noticeably blacker than cows. Jersey cattle are a good example of a breed showing wide variation in *Agouti* locus patterns. Jerseys are born tan, and develop blackness to varying degrees as they age. Bulls tend to be blacker than cows, so that while both sexes have the entire range of colors, uniformly tan bulls are rare and nearly black cows are likewise rare.

In many breeds the colors resulting from the combinations of black and tan typical of the *Agouti* locus are called "brown." This is especially the case when the tan has a relatively dark shade. Using "brown" in this sense can easily be confusing because "brown" more usually refers to the color of altered eumelanin as is the case in sheep and goats. The color of brown eumelanin in cattle is usually called "dun." In any case, "brown" is used by many people to describe cattle with combinations of tan and black because then the basic colors of cattle can be referred to as black, red, or "brown."

8.3.1 Brindle

Brindle in cattle is a pattern of tan and black stripes (Figure 8.8). Brindle is dominant to its absence. Brindle varies in the degree of black that is present, so that brindle areas can appear as tan with black stripes, or black with tan stripes. Brindle in Normande cattle is due to a specific *Agouti* locus allele, A^{Br}. It is dominant over all other alleles at this locus. The

Figure 8.8 Brindle cattle have stripes of tan and black. Animals with white spotting tend to split out the tan and black areas into more discrete spots, as is evident in this Florida Cracker heifer.

entire range of brindle phenotypes is confusing, and it is tempting to think that the final brindle pattern varies depending on background patterns consistent with other *Agouti* locus alleles. Brindling is often limited to the head and shoulders, but equally it can occur over the entire body. This distribution suggests that expression might depend on the underlying *Agouti* genotype and the relative extent of black coded by that. Many breeds and breed crosses have brindling over the entire body which suggests that the A^{Br} allele is able to provide the black that is then reorganized into the distinctive pattern of stripes.

Brindle in cattle is an example of a "mosaic" pattern where the tan-based and black-based areas are somewhat randomly developed. The consequence is that when brindle is combined with any sort of white spotting the black and tan stripes separate out into larger areas rather than narrow stripes. These areas are usually reasonably distinct, and often lack the striped character of most unspotted brindles. This trend for the colors to separate is most evident near white spots, and tends to diminish away from them. The tendency for the two pigments to separate from one another is especially true when brindling is combined with some of the dramatic white spotting patterns with speckling. The result is often a combination of distinctly tan spots and distinctly black spots over a white background. On many of these cattle the usual striped character of brindling is difficult to discern.

Brindle cannot be expressed on a uniformly red color, which makes sense in light of the mechanism of action of the *Extension* and *Agouti* loci. One of the best illustrations of this is the uniformly brindle calves that result from mating Hereford cattle (*red* at *Extension*, *brindle* at *Agouti*) with American Brahman cattle (*wild* at *Extension*, and a non-brindle allele at *Agouti*). The calves pick up *brindle* (A^{Br}) from the Hereford, even though the Hereford cannot express the pattern due to the recessive E^e genotype of the breed. The Brahman parent contributes the *wild* allele at *Extension* as well as other *Agouti* alleles. This cross consistently produces brindle calves even though neither parent is brindle.

Brindle cattle have a few perplexing phenomena. A few animals that are heterozygous for *dominant black* may be dark brindle. Another odd brindle and black phenotype occurs in some families of De Lidia (fighting) cattle in Spain. In these animals the barrel is brindle, and the head, shoulders, legs, belly, and rear are black. This is unexpected in most brindles, because in this example the brindle is not involving the black regions, only the tan ones. These animals generally have a completely black muzzle rather than the pale tan muzzle of nearly all *Agouti* locus patterns, which is another perplexing detail, and may point to a more complicated control of color than the usual alleles at *Extension* and *Agouti*.

8.4 Dilution

The dilution of color in cattle results in pale colors, and is important in arriving at the final distinctive basic color of several breeds. A few important loci are well documented as having roles in dilution, even though some specific dilution alleles are not yet assigned to a specific locus. These unassigned alleles are well documented by segregation data, so their mode of inheritance is known, even if their locus of residence is not in some cases. The multiple alleles responsible for dilution each generally occur in several different breeds. All of them are relatively frequent in cattle, in contrast to the situation in goats, sheep, llamas, and alpacas where dilution alleles are both few in number and rare in frequency.

Figure 8.9 Dun cattle in the Dexter breed have a recessive *Brown* locus allele.

An allele at the *Brown* locus changes black eumelanin to a flat, chocolate brown. It also lightens skin and eyes (Figure 8.9). A recessive allele, *brown*, is responsible for this change. The resulting color is called "dun" in Dexter cattle. Dun is a medium brown shade that is generally uniform over the animal, especially if *dominant black* is present. The *brown* allele only lightens eumelanin and does not affect tan at all, although the skin and eye changes would still be evident to a careful observer. This *brown* allele is the only source of dun in Dexter cattle.

The dilution typical of Charolais cattle affects both pheomelanin and eumelanin. The *charolais dilution* allele that accomplishes this is incompletely dominant. Homozygotes are nearly white (Figure 8.10). In heterozygous cattle tan or red is lightened to yellow (Figure 8.11), and black is lightened to a medium smoky grey color. The *charolais dilution* allele resides at the *Silver* locus.

Figure 8.10 The *charolais dilution* allele is named after the Charolais breed, but occurs in several other breeds as well. Homozygotes are white or nearly white (photo by J. Beranger).

Figure 8.11 The *charolais dilution* allele dilutes red to yellow in heterozygotes. It is a common source of yellow color in the Griffin bloodline of Pineywoods cattle (photo by J. Beranger).

Another allele at the *Silver* locus is also incompletely dominant but has only moderate potential to dilute colors when compared to the *charolais dilution* allele. This other allele accounts for the pale color of Simmental, Highland, Galloway, and Belted Galloway cattle. The *simmental dilution* allele usually has minimal effect on pheomelanin, with a slight change to a light red or dark gold (Figure 8.12). It has a more extreme effect on eumelanin which is reduced to a smoky grey (Figure 8.13). Homozygotes are paler than heterozygotes, but always retain a much darker color than homozygotes of *charolais dilution*.

Figure 8.12 Most Simmental cattle are a light red as a result of their unique *simmental dilution* allele. Some cattle of the breed lack this diluter, and are fully intense red.

Figure 8.13 On a black background color the result of *simmental dilution* is a light smoky color that is typical of dun Belted Galloways.

Most dun Galloway cattle are caused by the *simmental dilution* allele. Occasional dun Galloways are produced following matings of black cattle, and this is consistent with a recessive phenomenon. The mode of inheritance and final color are similar to the *brown* allele present in Dexters, and are likely due to that allele. Having two distinct types of dun that are visually similar in one breed can potentially lead to surprising results in offspring colors.

The cause of the final color of Brown Swiss cattle is also consistent with the action of a dominant dilution allele (Figure 8.14). The final dilution of black pigment in this breed is only moderate, and this level of dilution is present in homozygotes. For these reasons it is similar to the effects of *simmental dilution*. In Brown Swiss cattle this dilution is superimposed over various *Agouti* patterns rather than the solid color caused by the *dominant black* background of Galloways. The difference in background color affects the final phenotype even though the color and character of the dun regions are similar.

Other breeds and strains of cattle have candidates for yet other dilutions. Several of these segregate as dominant alleles, passing reliably from generation to generation without

Figure 8.14 These two Brown Swiss cows show the effects of dominant dilution on black areas, and also the range of the final effect in a breed that expresses the *Agouti* patterns and has relatively pale tan areas.

any breaks. These include "Larson Blue" Holsteins, which are a pale, greyish color that is diluted from black. Murray Grey cattle descend from an initial Shorthorn x Angus cross, and they are all a dilute grey color. This color likely comes from the *simmental dilution* allele, and must have been contributed by the Shorthorn ancestor. How this incompletely dominant allele could be provided by a breed not known for pale color is something of a mystery. The mostly likely explanation is that the effect of *simmental dilution* on red cattle is often minimal and can escape notice. The Shorthorn ancestor could have had this allele, and yet could have had an acceptably dark enough shade of red color such that breeders accepted the animal without reservation. While the effects of *simmental dilution* on red can be subtle, its presence on a black background color is a very obvious blue-grey color, and Murray Grey cattle are homozygous for the dilution. Their color is relatively dark and could never be mistaken for something as extreme as *charolais dilution*.

Some dilute calves have abnormal hair coats. This includes a sparse or absent tail switch, and the phenotype is therefore called "rat tailed." The color is a dark grey dilution of black that seems to not affect red or white areas of the coat. The hair in the affected regions is short and curly. These calves usually occur from Simmental parents mated to black mates. This gene is probably dominant, but is only expressed on a black background color so that the Simmentals, with their red color, never express it. The gene appears to be rare in Simmentals, because most of the many grey dun hybrids that the breed produces are normal and lack the rat tail or the other hair changes associated with this phenotype. Rat-tailed calves have a lower tolerance to the cold than normal calves, so breeders try to avoid them.

8.5 Albino and "Near Albino" Colors

The *Albino* locus in cattle likely has a few different mutations. One of these leads to albino animals that are nearly white and have unpigmented eyes. This phenotype has been described as being due to recessive alleles in Beef Shorthorn, Charolais, Guernsey, Brown Mountain (Swiss and Austrian), Murboden, and Simmental cattle. Whether the allele is identical in all breeds is uncertain.

A different form of incomplete albinism, at an undetermined locus, has been reported as a recessive allele in Holsteins. This allele reduces black to a cream color leading to a "ghost" pattern on the normally spotted animals of the breed. The eye is blue-grey. A similar color occurs in Brown Mountain cattle, but with light blue-green eyes. These may all be similar or identical genes. A separate dominant gene is responsible for a very pale phenotype with a few pale spots. The eyes are usually blue centrally surrounded by a very pale iris. These are called wall-eyed albinos. The Chediak-Higashi syndrome of cattle involves dilution of color to a very pale shade with light grey eyes. This phenotype has moderately defective white blood cells that perform poorly. As a result these cattle are not as viable as normal cattle, so the color is of more theoretical than practical importance. This is due to a recessive allele.

A consistent phenotype of dilute tan areas with fully expressed black areas has been postulated to be due to a *chinchilla* allele at *Albino* (*Tyr* or *Tyrosinase* locus of the mouse). The allele has various characteristics that suggest that it most likely resides at a different locus. The allele alters the *Agouti* patterns to be a combination of cream and black instead of the

Figure 8.15 This pair of zebu cows includes one that never transformed from her tan birth coat to the usual silver adult coat.

more expected tan and black, or red and black, typical of unmodified *Agouti* locus patterns. Cattle with this sort of dilution are common in zebu breeds, including the American Brahman breed of the USA, but also the Gujerat (Guzerá) and Nellore breeds. The same pattern of pigmentation is also typical of Italian breeds such as the Chianina.

In these breeds the calves are born a medium to pale tan. The black areas then develop and the tan areas become silver. Breeders usually refer to this result as "grey" or "silver" (Figure 8.15). In the Chianina and related breeds the black areas are minimal, and these cattle are usually simply referred to as "white." The zebu breeds with this form of dilution serve especially well as examples of *Agouti* locus patterns because of the contrast between silver (diluted tan) and black areas. All of the cattle of several zebu breeds have silver tan areas. Others, such as the American Brahman, retain both the silver and red extremes of tan pigmentation, as well as all of the intermediates. This allows for more variation in the final color phenotype of the breed than is typical of most other zebu breeds.

The development of color is consistent across the "silver" breeds. They progress from a tan birth coat to a final cream and black color. This trajectory is opposite to that expected from a chinchilla-like allele because such alleles in other species produce animals that are generally incapable of forming dark pheomelanin. The tan birth coats of these breeds should therefore be impossible with a chinchilla-like mode of action. Whatever the actual locus involved, the dilution in these breeds is consistent with a recessive mode of inheritance. The dilution tends to be lost in heterozygotes with most other breeds.

8.6 White Spotting

White spotting is rare or non-existent in most wild mammals, including the aurochs ancestor of cattle which is now extinct. The widespread occurrence among domesticated cattle is interesting, and at least one study suggests that the color of finely mottled, speckled, or spotted cattle disorients bloodsucking flies to the extent that they fly away to annoy other less flashy targets. Spotted and speckled cattle therefore have a significant advantage in environments with flies. Spotted cattle are also attractive to their human owners, and both the fly resistance and the visual appeal have probably contributed to the widespread popularity of spotting patterns.

White spotting patterns are common in several breeds of cattle. Specific patterns have become an essential part of breed identity for many breeds. Some breed associations, such as the ones for the Holstein, Guernsey, Hereford, and English Longhorn, have used specific white spotting patterns to distinctively stamp their breed. Other breed associations, especially those for landraces, allow cattle to vary widely in both color and white spotting patterns. Many interesting and beautiful patterns persist in the landrace breeds. Unfortunately for those interested in characterizing color genetics, many of the spotted patterns in landraces are produced by combinations of multiple individual patterns, and the specific individual patterns can be difficult to identify in these situations. Several of the spotting patterns, including *spotted, colorsided, hereford,* and *pinzgauer* all reside at the same locus. As a practical issue this means that not all combinations are possible. Only two of these can be combined in any one animal, despite the many choices at a single locus. The *Spotting* (or *KIT*) locus is the site of these alleles.

Many of the spotting patterns on cattle are roughly symmetrical from one side of the animal to the other, which can provide an important clue to the genetic mechanisms producing these. The knowledge that nearly all of the symmetrical patterns are inherited as dominant alleles can be particularly helpful to breeders managing these patterns in populations. The fine details of inheritance, as well as the genetic location of many of these alleles, are as yet undetermined. Accurate phenotypic identification is a good first step in accomplishing those studies, which is the reason for discussing these patterns in sufficient detail to focus on the differences among them.

The white spotting patterns in cattle follow the same rules that hold true across all species. Each pattern varies in expression from minimally white to maximally white. This trajectory is usually characteristic for each individual pattern, and nearly all animals can be accurately classified. Many of the patterns vary over a relatively narrow range, so that the "least white" has enough white to be classed as spotted while the "most white" retains tell-tale signs of the patterns present.

The one exception to the general rule of distinctions between the patterns is the "white park" pattern that derives its name from the White Park breed. This pattern is a white coat with pigment remaining on the ears, muzzle, feet and teats and around the eyes. This pattern is usually described as "white with colored points," and is the most extensively white endpoint of a few different patterns. This one pattern, therefore, cannot always be related to the expression of only one single identified allele. Evaluating an entire herd of cattle, rather than just one animal with the white park pattern, can provide some clues as to the specific allele involved. Fortunately the most common allele causing the white park pattern is the *colorsided* allele, and in most situations this allele is a prime candidate. However, other alleles can also produce this pattern, and these others may be present in some populations. These include *roan, morucha, colorsided,* and perhaps *fisheagle* as well as *salinero*. The fine details of the white park pattern are therefore confusing because it is the most extensively white expression of several patterns.

Table 8.2 The several white spotting patterns of cattle can be sorted into different major groups by the character of the white areas.

clear, crisp white	speckled	roan	modifications
spotted	*colorsided*	*roan*	*brockle*
hereford	*fisheagle*	*morucha*	*ticking*
simmental	*spitting cobra*	*salinero*	*smudge*
pinzgauer	*agricola*	*pedi*	
belted	*bororo*		
wading	*speckled sides*		
white sides			

White spotting patterns can be broken down into a few major groups, as outlined in Table 8.2. The patterns within a group share certain characteristics, but each also has distinctions that clearly identify them as separate. The major groups are those with crisp white areas, those with speckled white areas, and those with roaning. Modifications of these patterns are important, and put colored spots or hairs back into the white areas.

A fairly consistent pattern of white spotting has white on the legs, tail switch, and forehead (Figure 8.16). This can be a minimal expression of the *spotted* allele at the *Spotting* locus, but can also occur from a handful of other genetic mechanisms. This combination of white occurs fairly commonly across nearly all domesticated species. In nearly every species this specific phenotype can be produced by a handful of different genetic mechanisms.

The *Spotting* locus (also called *KIT*) has a recessive allele, *spotted*, which is responsible for irregular white spots (Figures 8.17 and 8.18). The minimal pattern is usually white on the forehead, legs, and tail switch. More extensively white expressions have irregular spots on the body. This is a common pattern in many breeds. Some *spotted* animals have pigmented lower legs, which likely comes from a separate genetic instruction (Figure 8.18). Breeds vary in uniformity for the *spotted* allele. It is consistently present in Holsteins, Guernseys, and Ayrshires. It is common in Shorthorns, and somewhat rare in Jerseys.

Figure 8.16 This Criollo Patagónico calf has a pattern of white on the face, feet, and tail that is likely a minimal expression of *spotted*.

Figure 8.17 Many breeds, including the Criollo Patagónico, have *spotted* individuals.

The extent of white on *spotted* animals is controlled by modification at other loci. The extent of spotting has a high heritability, and breeders can therefore select cattle to be either more extensively, or less extensively, white if they desire to do so. The variation in extent of white is more than cosmetic and has implications for adaptation to various environments. This has been best documented in the Holstein breed, where the whiter individuals do better than the blacker individuals when they are in hot climates or have intense solar radiation. The opposite trend is documented less well, but some observers note that the blacker animals do better than whiter ones in colder environments. These differences are likely due to the two extremes having differential absorption of radiant heat from sunshine.

Figure 8.18 This Nguni cow is *spotted*, but has dark legs. The exact cause of this combination is uncertain. The small black spots in some white areas suggest that *brockle* may be involved, although this cow has only a minimal expression of that modification.

Figure 8.19 This cow has the *hereford* pattern in the medium range of expression.

Figure 8.20 This cow has a more maximally expressed *hereford* pattern.

The Hereford pattern of a red body with white face, legs, and tail switch is due to a dominant allele, *hereford*, that resides at the *Spotting* locus. Hereford cattle all have this pattern. Heterozygotes (Hereford crossbreds) have less white than homozygotes (Figures 8.19, 8.20, and 8.21). This could be influenced through two or three different mechanisms.

Figure 8.21 This cow is heterozygous for the *hereford* pattern, and is also black. The combination usually leads to a fairly minimal expression of the pattern.

Figure 8.22 The pattern of Simmental cattle is caused by an allele distinct from Herefords, even though they both produce a white face.

It might well be that the Hereford pattern is more extensive in homozygotes as an inherent characteristic of the action of the allele. Equally, it might be that the Hereford breed has simply had selection for relatively extensive expression of the pattern, and that the modifiers causing extensive markings are disrupted in crossbreds. Many heterozygous animals are crossbreds, and specifically are crossbreds with black breeds such as the Angus. That specific cross is popular, resulting in a uniform crop of "black baldy" calves that are highly regarded for adaptation, ruggedness, and overall production. The black background color of the crossbreds may well be influencing the expression of the white spotting in a negative direction, which can contribute to the final phenotype of less extensive white regions when compared to purebred Herefords with their red background color.

The *hereford* allele occurs in other breeds, including central Asian breeds, the Groningen breed from the Netherlands, and the Pampa Chaqueño of Paraguay. In those breeds it is less consistent in expression than it is in the Hereford. The more consistent range of expression in Herefords is probably due to long-term selection for appropriate modifiers in the Hereford breed.

The white face does not come without consequences. Animals with unpigmented eyelids have an increased risk of developing squamous cell carcinomas ("cancer eye") than those with pigmented eyelids. On the other hand, cattle with white faces have fewer horn flies than those with colored faces.

Simmental cattle have a spotting pattern that is superficially similar to the *hereford* pattern because they consistently have a white face. The *simmental* allele resides at the *Spotting* locus. The *simmental* white face is usually less extensive, and many cattle with the *simmental* allele have white spotting on the body (Figure 8.22). This is a dominant allele that is distinct from the *hereford* allele. Cattle with *simmental* are easily confused with *hereford* if they lack spotting on the body, despite the fact that these two similar patterns are each caused by a different allele.

Figure 8.30 The most extensively white expression of the *colorsided* pattern is the white park pattern and bears the name of the breed in which it is a defining characteristic (photo by J. Beranger).

another chromosome in some breeds. The pattern has an interesting worldwide distribution, even being present in domestic yaks and gayals.

The final appearance of the *colorsided* pattern varies from breed to breed, even though it is due to a single well-documented genetic mutation. This is likely due to modifiers that adjust the extent of white to be either more extensive or less extensive. Some breeds, such as English Longhorn, tend to converge on the least white extreme even though the cattle are all homozygous for the allele. Others, like Randall Linebacks, vary over a wide range but never quite reach the white park extreme. Instead they retain at least some color on the sides even though that might only be as flecked or roan areas. White Galloways that result from the *colorsided* pattern vary from the usual white park pattern to one that has colored flecks on the sides.

Notably, the White Park breed that lends its name to the extremely white version are indeed white with only minor color remaining on the ears, nose, feet, and teats, as well as around the eyes. White Park animals vary in genetic constitution, some being heterozygous and others homozygous. Occasionally animals are therefore born that lack the mutation and are nonspotted or nearly so. The heterozygous animals are usually as white as their herd mates, although many homozygous White Park animals have minimal pigmentation on the ears. This contrasts to the more classic pattern with fully pigmented ears.

A few breeders of Florida Cracker cattle have noticed that heterozygous animals tend to all have obvious and easily identifiable expressions of the *colorsided* pattern. Homozygotes in that breed tend to have the white park pattern. This neat and predictable genetic mechanism unfortunately does not hold for all breeds. In breeds such as the Randall Lineback and

Figure 8.31 Nguni cattle with the *fisheagle* allele have white on the head and rear and minimal flecking on the body.

the English Longhorn all animals are homozygous, and yet the range of expression includes some individuals with minimal white and few or none that are maximally white.

The *fisheagle* pattern of Nguni cattle is similar to *colorsided*, but on cattle with minimal expression the flecking occurs on the head and rear, tending to spare the middle of the body as seen in Figure 8.31. The head is usually extensively white. The ideal pattern affects nearly the entire head, just like the bird species that lends its name to the pattern.

Another distinct white flecking pattern of Nguni cattle is the *spitting cobra* pattern (Figure 8.32). This pattern has white flecking on the rear and throatlatch, and varies considerably in the extent of the white. The most extensive manifestation is likely to be the white park pattern.

Within the Agricola strain of Pineywoods cattle, and also in other breeds, is a pattern of

Figure 8.32 The *spitting cobra* pattern of Nguni cattle has white flecks at the throat and on the rump.

Figure 8.33 The *agricola* pattern is speckled fairly uniformly over this Pineywoods cow.

flecking and mottling that is fairly evenly distributed over the entire animal (Figure 8.33). The maximal extent of this *agricola* pattern is likely to be the white park pattern, at which point it could be easily confused with others. In the middle and minimal grades of expression the uniform distribution of the pattern helps to distinguish it from other patterns.

The Florida Cracker and a few other breeds have a rare pattern of white spotting that puts white on the lower half of the body, with a transition zone between the white and colored areas that is made of small spots of color and white (Figure 8.34). This *speckled sides* pattern passes from parent to offspring fairly regularly, indicating a likely dominant mode of inheritance. The flecked boundary of the white distinguishes this from the *white sides* pattern. The pigmented legs distinguish it from the *wading* pattern.

Among the generally red Mbororo cattle of Africa are a few that have fairly evenly distributed small white spots that tend to be circular and up to about 3 cm in diameter (Figure 8.35). These spots are remarkably evenly distributed across the entire animal instead of the more regionally concentrated distribution that is a characteristic of most other spotting patterns. No studies have demonstrated how this pattern is inherited.

Figure 8.34 Some Florida Cracker cattle have the distinctive *speckled sides* pattern of white on the sides that intergrades to the body through a speckled region.

Figure 8.35 The uniform speckling of this Mbororo cow is different from other white spotting patterns (© Werner Lampert GmbH, photo by Fabrice Romain Monteiro).

8.6.1 Roan

"Roan" is often used in a generic sense to describe any relatively even mixture of white and colored hairs. In many cases it is also used to describe finely flecked or speckled animals. The Shorthorn breed has many roan animals. The pattern in this breed is the usual color pattern referred to by using "roan" as an unmodified word. This pattern usually varies in intensity across the body of the animal, so that some areas are lighter, and some are darker (Figures 8.36 and 8.37). The *roan* pattern is incompletely dominant. Heterozygotes are roan while homozygotes are white or nearly so. Homozygotes usually retain color on the ears, and sometimes on the feet, and muzzle and around the eyes, in a pattern similar to the

Figure 8.36 Shorthorn cattle with the *roan* pattern often have an uneven distribution of roan, red, and white areas (photo by J. Beranger).

Figure 8.37 This Shorthorn cow has a typical expression of the *roan* allele (photo by J. Schallberger).

Figure 8.38 White Shorthorn cattle are homozygous for *roan*. They usually retain pigment on or inside the ears, or on other peripheral body regions such as the feet (photo by J. Beranger).

white park pattern (Figure 8.38). The mutation for *roan* occurs at the *Steel* locus which is also known as *Mast Cell Growth Factor*.

In Spanish and African breeds a distinct type of uniform roan occurs (Figure 8.39). It is nearly uniform throughout the Morucha breed from Spain. The *morucha* pattern varies from dark to light, and usually has a uniform mixture of colored and white hairs over the entire animal. The white hairs spare the feet, muzzle, and ears, around the eyes, and around the anus or vulva. The most extensive pattern therefore results in the white park pattern: a

Figure 8.39 The *morucha* pattern of uniform roan can be seen in a number of breeds, such as this Criollo Patagónico cow.

Figure 8.40 The *salinero* pattern of roan varies, but nearly always involves the rump and the heart girth as evident in this Mertolengo cow (photo by the Associação de Creadores de Bovinos Mertolengos).

white animal with colored points. Breeders of Morucha cattle carefully monitor the extent of white, avoiding the whitest extremes by occasionally crossing to the black cattle of the breed to darken the shade of *morucha* roan.

Another pattern common in Iberian breeds is called "salinero" in Spanish, or salt worker (Figure 8.40). These cattle, when minimally marked, have flecked roaning on their rears. The *salinero* pattern then tends to go to the shoulder, heart girth, and head. The most extensively white animals also have roaning and flecking on the barrel, neck, and head. On extensively marked animals the rear and heart girth are usually whiter than the barrel. The most extensive pattern is the white park pattern. The *salinero* pattern varies from distinctly roan to more speckled in the Mertolengo breed of Portugal as well as the Texas Longhorn.

Some animals have roan heads, with the roan sparing the ears, eyes, and muzzle. This pattern is rare, and appears distinct from the other types of roaning.

The *pedi* pattern comes from a group of Nguni cattle by that name. These cattle are highly esteemed by their owners for their large size and high growth rate and also for their beautiful color pattern. Most *pedi* cattle have a subtle interplay of darker and lighter roan areas over the body that is over pigmented skin (Figures 8.41 and 8.42). They also tend to have at least a few isolated round spots that are of the dark background coat color and are scattered somewhat randomly. In the Nguni it is difficult to tease out which elements of this pattern are limited to the gene causing *pedi*, because the pattern is commonly combined with other patterns. Some *pedi* cattle have darker heads and legs, and it is tempting to suggest that this might be the "uncombined" pattern, and that the more common ones

Figure 8.41 The *pedi* pattern is one of the most distinctive of the Nguni patterns. The irregular roaning and random dark spots are both evident on this cow.

Figure 8.42 The same pattern typical of *pedi* cattle is seen on some Gyr cattle. The scattered dark areas are typically the same as the background color on the body region bearing the spots.

with lighter legs and heads may also have the *fisheagle* or *colorsided* patterns. A pattern similar or identical to *pedi* occurs in some Gyr cattle.

8.6.2 Combinations of White Spotting Patterns

Various combinations of white spotting patterns are possible on cattle with the exception being that those residing at a single locus can never include more than two different patterns. In most standardized breeds the combinations are fairly rare, largely because breeds tend to be standardized around a single color along with a single white spotting pattern. The phenotypes in most breeds are therefore usually limited to a single spotting pattern, caused by a single allele. For example, Holsteins are black and *spotted*, Guernseys are golden red and *spotted*, Pinzgauers are red and have the *pinzgauer* allele. Jerseys, in contrast, vary in color and while usually unspotted, can also be *spotted*.

A few combinations are not intuitive at all. Some Holsteins have very extensively white heads, with color around the eyes and on the ears as well as small round spots on the middle of the sides. This is likely a combination of *spotted* and *colorsided*, although the roan areas typical of *colorsided* are rarely, if ever, present in the combination. The *pinzgauer* pattern also occasionally breaks up into small round spots on the sides, and is likely a reflection of some combination of alleles (Figure 8.43). Many Nguni cattle have a pattern that is likely the result of combining *colorsided* with *fisheagle* (Figure 8.44). They have large areas of color

Figure 8.43 This Pineywoods cow is likely a combination of *pinzgauer* and another allele, all on a background that is yellow from a dilution of red (photo by J. Beranger).

on their sides, but extensive white areas on most other regions. A few patterns, especially in landraces, can be nearly impossible to decipher. Some of these may be independent single patterns that are as yet uncharacterized (Figure 8.45).

Cattle with combinations of multiple spotting patterns are often more extensively white than expected from a simple addition of the two patterns. This implies that the patterns may be additive when they combine, leading to an extension of white that comes from each pattern assisting the other in extending the white at the expense of pigmented areas.

Figure 8.44 Nguni cattle with dark side panels that are otherwise flecked or roan probably have a combination of patterns.

Figure 8.45 This Nguni calf has a striking pattern that is likely a combination of at least two others, but that also might be an uncharacterized single pattern.

8.7 Modifications of White Spotting

A few modifications of white spotting occur in cattle. These add color back into areas that should be white if the spotting patterns are unmodified. The resulting final color patterns can be difficult to decipher in some situations, and include many striking and unique patterns. They are especially common in breeds that vary widely in color.

Some cattle have small round spots in white areas. These are present at birth, and are similar to the *brockle* pattern in sheep and goats (Figure 8.46). This effect varies from extensive to minimal. In most cases *brockle* is the cause of the "colored back" pattern, when combined with either *pinzgauer* or *colorsided* spotting. These animals have small colored spots, but these tend to coalesce into larger spots over the top of the animal, leading to an irregular, colored stripe along the back. The combination of the *pinzgauer* pattern and *brockle* is typical of Bordelaise cattle. Not all Bordelaise cattle are homozygous for *brockle*, though, so occasional calves are produced that lack *brockle* and therefore have the unmodified *pinzgauer* pattern with its clean white regions. Brockled animals have irregular color in areas that should be white by alleles of white spotting (Figure 8.47). This is something of a

Figure 8.46 This *brockle* Pineywoods has the typical spots and dark backline, all on a background *pinzgauer* pattern (photo by J. Beranger).

Figure 8.47 Brockling adds color into white areas, especially the white face of Herefords, as seen in this Criollo Patagónico bull.

theoretical conundrum, but is widely reported. For example, Ayrshire cattle with dark legs likely have this effect, as do crossbred Hereford cattle with colored spots in the white face.

The *ticking* modification causes small flecks of color to grow into white areas (Figure 8.48). This is not present at birth, but usually begins no later than one year of age. The results can vary from extensive to minimal. When extensive it sometimes gives a roan appearance. In many animals the areas of white that are most distant from the colored area become ticked, but the areas immediately adjacent to colored areas do not become ticked.

Figure 8.48 Small spots of color in white areas are the result of *ticking*, as seen in this Criollo Saavadreño cow.

Figure 8.49 The difference between *smudge* and *ticking* is the tendency for *smudge* to be individually colored hairs which are added into the white areas. This Pineywoods cow is a complicated combination of *pinzgauer*, *brockle*, and *smudge* (photo by J. Beranger).

This has the effect of leaving a white rim around the colored areas with a more speckled or roan area inside the white areas.

The *smudge* modification resembles *ticking* because dark hairs grow into white areas (Figure 8.49). In this case, the colored hairs do not coalesce into spots, so the result is more roan than spotted. The extent of *smudge* varies considerably, leading to a wide range of final appearances. It is most likely due to a dominant allele.

CHAPTER 9

Putting Knowledge to Work: Cattle

Many breed associations for cattle have adopted a tight relationship of coat color with breed identity. This leaves little room for the breeders of most breeds to explore the development of new color variations. In some breeds, however, enough variation does exist that interested breeders can explore color variation and the production of certain combinations. Additionally, some breeders take advantage of the details of color genetics to assure that the breed combinations that they use are evident for all to see in the cattle they produce.

Genetic tests are available for some of the alleles that produce color variation in cattle. Genetic tests are generally developed when researchers see the potential for enough testing to generate a reasonable profit. Test availability has varied over time, and fortunately this has usually been in the direction of an increasing number of tests becoming available. The currently available tests for cattle include:

1. *Extension* locus alleles (*dominant black, brindle, wild, red, red charlie*)
2. *dominant red* of Holsteins
3. *charolais dilution*
4. *simmental dilution*
5. *brown.*

The presence of several tests can help breeders interested in specific outcomes in their herds, although for most breeders color is less important than other production-related characteristics. As more and more testing is done, breeders also find occasional results that seem to be contradictory to the observed final phenotype of the animal that was tested. These results can be very perplexing. For example, a test result might indicate that an animal is black when a breeder sent samples from a very red animal. This type of discrepancy reflects the complicated nature of genetics. The tests are developed to detect variations that are known and documented. As more and more animals are tested from an increased variety of breeds, it is also increasingly likely that breeders will encounter novel genetic mechanisms for some variants. Existing tests will not be able to pick up these newly encountered mechanisms.

Figure 9.1 Red cattle, such as this white-faced Pampa Chaqueño cow, can arise from several different genetic mechanisms.

Red cattle are a good example of this (Figure 9.1). A fairly decent red phenotype can be produced from any of four genetic mechanisms. At *Extension* this includes *red*, and *red charlie*, but can also include *wild* if black markings are minimal and pheomelanin is fully saturated. The *dominant red* allele can also cause a red phenotype that is only minimally distinguishable from these other types. The existence of *dominant red* and *red charlie* was only suspected when previous DNA tests for color indicated that the red animals being tested should be black. This alerted researchers to the fact that some new mechanism was operating, and so they set out to document it. New alleles were found as a result of their efforts.

Dun is another final color that can be reached by many different pathways. At least three of these are documented (*charolais dilution*, *simmental dilution*, and *brown*). No doubt additional examples of mechanisms are waiting to be discovered, such as the specific allele that causes the rat-tailed phenotype or the Larson Blue Holsteins.

9.1 Solid Colors

Strategies for manipulating the various solid colors of cattle are fairly straightforward. If the goal is the production of cattle that are obviously and darkly colored, then breeders should avoid mating dilute cattle together, especially those with *charolais dilution* (Figure 9.2). An alternate strategy of only mating a dun or gold animal to red or black mates allows for the production of both dark and medium colored calves, but completely avoids the nearly white ones. The resulting calf crop is variable, but specifically excludes the nearly white calves that may not be desired.

Figure 9.2 This yellow Pineywoods cow was mated to a yellow bull to produce a cream-colored calf. These two also have *pinzgauer* spotting.

In contrast, if a breeder desired a uniform calf crop that includes nothing but gold calves, the best strategy is to mate a white Charolais bull to red cows. If the goal were dun or grey calves, then the cows should be black. This strategy assures that the calves are all heterozygous for *charolais dilution* and are therefore a reasonably uniform color. The calves will pass along the allele only about 50% of the time, but they themselves will all be the same color.

Managing the coat color of Galloways can be a slightly more complicated endeavor. The main choices are black, red, and dun. The difference between black and red animals is the usual straightforward control at the *Extension* locus. Dun, in this case, is generally due to *simmental dilution*, and this affects black more than it does red. The potential snag here is that many red cattle that are heterozygous for the diluter are so minimally affected as to be misclassified. They can therefore produce dun calves from black mates, which can be a surprising result.

A recent trend in the USA finds many cattle breeders changing color preferences to include black, either with or without white spotting. This occurs in breeds that historically were not black, such as the Simmental. Original Simmentals are red with a white face and body spots. To introduce black into the breed, the breeders used upgrading. Upgrading is the sequential use of purebred sires over crossbred and part-bred cows, until after several generations the genetics of the purebred breed predominate. In this case, the initial cross is likely to be a Simmental bull over Angus cows. These calves are 50% Simmental, and are all black heterozygotes that also have the white face and body spots of the Simmental. If the heifer calves are taken back to a different Simmental bull the result is calves that are 75% Simmental, and half are black. Half have the full Simmental pattern of white, half are only heterozygous for that pattern. Over succeeding generations it is possible to eventually end up with a nearly pure Simmental that is black and has minimal white markings, by consistently selecting out the black calves with minimal white marks at each generational step.

At the point along the upgraded generations that the cattle can be considered Simmental, they can be mated to one another instead of to a Simmental bull. The cattle involved in this step will all be heterozygous $E^E E^e$ and $Sp^S S^+$. When mated to one another they will produce about 75% black animals. Of those black animals, about a third will lack white markings. The final yield of "black Simmentals" is then about 18%. Despite their uniform black color, these calves otherwise have largely Simmental genetics. They can be mated more broadly in the breed to produce increased numbers of black animals that are heterozygous for the white pattern, with the eventual goal of increasing the numbers of solid black animals if that is desired.

This specific example shows the ease with which a dominant allele can be introduced into a breed, which in this case is *dominant black*. Even over many generations of upgrading it is possible to accurately track the *dominant black* allele and to retain it as the final goal. In contrast, ridding the final product of a dominant allele that is present in the target breed is more difficult. In this case, the *simmental* spotting pattern is dominant, but fortunately has at least some incompletely dominant tendencies. This allows a breeder to preferentially retain heterozygotes by retaining the calves with the least amount of white. The final goal is the production of cattle that are homozygous for the introduced *dominant black* allele, but also homozygous for *wild* at *Spotting*. This last step must be delayed until the final upgraded products can be mated to one another, because the Simmental parents keep introducing the *simmental* allele for spotting at each generational step. If an upgrading project desired to eliminate a dominant allele with no incomplete dominant tendencies, then the result would be difficult to attain and would likely require either test mating, or a good genetic test in order for heterozygotes to be retained at each generation.

9.2 Spotted Colors

In general, spotting patterns are uniform throughout standardized breeds so there is little variation to manipulate. Some breeds such as the Holstein are uniform for some patterns and minimally variable for others. Holsteins always have *spotted* but a few also have *color-sided* as well as *hereford*. The interactions of the rare alleles with the common *spotted* pattern provide for interesting, if subtle, variations in the breed. Holstein breeders focus on milk production more than color, so the minor contributions of the rarer spotting alleles are generally not considered.

Patterns of white spotting are allowed a large range of variation in most landrace breeds. In the selection setting of many landraces, breeders try to avoid animals that are largely white. To avoid the production of extensively white calves it is prudent to mate the cattle that have more white to solid-colored mates. For several of the roan or speckled patterns the heterozygotes tend to be more dramatically marked than the homozygotes, although this varies from breed to breed and strain to strain. In some breeds, and for some spotting alleles, the homozygotes are likely to converge on the white park pattern. Avoiding the production of homozygotes therefore reduces the chance of a white park calf. This is a general trend, though, and not an absolute rule.

Adding two or more white spotting patterns together can blur the distinctiveness of each one, and can produce animals that are predominantly white. The general trend in this instance is that the patterns have at least some tendency to boost the extent of white when they are added together. The final result is often more extensively white than would be the case by simply adding together the extent of white of each individual pattern that goes into the combination. Mating cattle with two or more patterns to solid-colored animals tends to split alleles for the patterns apart. The patterns then tend to manifest individually in a majority of the calves.

Figure 9.3 This Criollo Pilcomayo bull has a combination of *pinzgauer* and *spotted* resulting in an extensively white coat with round spots on the sides.

However, a few specific examples of combining patterns might well be desired by some breeders. The combination of the *pinzgauer* pattern with *spotted* can lead to attractive patterns with very round spots on the sides of the animal (Figure 9.3). The combination of *spotted* with *colorsided* can similarly lead to attractive round spots on the side of an animal. In these situations breeders can preferentially mate cattle with these combinations, producing a higher yield of the combination in the calf crop.

Some specific combination patterns are quite striking, highly desired by breeders, and present challenges if the goal is the consistent production of calves with the combination. Assuring that the required alleles are present in both parents will assure the highest level of production of the combination in the calves. The Bordelaise breed of France is an example. This breed has the *pinzgauer* pattern, which is due to a dominant allele. In addition it has the *brockle* modification of that pattern, which adds color down the back and small colored spots into the white areas (Figure 9.4). The combination is striking and distinctive, but not all cattle of the breed are homozygous for the *brockle* pattern. The result is that the breed also includes animals with clean white areas from the unmodified *pinzgauer* pattern. As long as at least one parent in each pairing is *brockle*, the calves stand at least a 50% chance of having the desired combination. Completely eliminating the production of the clean white variant would be a challenging endeavor especially in the absence of a genetic test for *brockle*. The temptation would be to eliminate all animals that lack *brockle*, and perhaps all heterozygotes as well. In a rare breed such as the Bordelaise it is wiser to include all animals

Figure 9.4 This Florida Cracker calf has the same color pattern as Bordelaise cattle, which comes from adding the *brockle* modification onto the *pinzgauer* pattern.

and slowly increase the frequency of *brockle*. This can be assured by the use of at least one *brockle* animal in each mating pair.

9.3 *Trademarking Crossbreds*

The general trend for many standardized breeds to have only a single color and a single spotting pattern has served breeders well over the years. This strategy has provided for easy and reliable breed identification by the casual observation of color. The advantages of this general trend are obvious for purebred breeders, but also give unique advantages to breeders producing crossbreds between specific breeds with the goal of cattle that perform well in specific environments. The crossbreds benefit from hybrid vigor, and also from the blending of the performance characteristics of the paternal and maternal breeds that went into the cross. Cows produced by certain specific crosses get a premium price in some markets because buyers anticipate the productive potential of these animals. Manipulating color genetics can easily put a distinctive stamp on these crossbreds that assures their easy identification. The usual technique for this has been to assure that each of the parental breeds is uniform for a dominant allele that will be expressed in the crossbred calves.

A few of these crossbreds have long histories of deliberate production. The "black baldy" is still produced in the USA, and is highly desired as a productive and well-adapted mother cow (Figure 9.5). The "baldy" portion of its name refers to the white ("bald") face it inherits from a Hereford parent, and "black" refers to the background color that the calves get from an Angus parent. These cows produce well in a variety of environments. The Hereford parent provides recessive *red*, and dominant *hereford*. The Angus parent provides *dominant black* and the recessive *wild* (or nonspotted). The resulting calves then uniformly express *dominant black* and *hereford*, a combination that is the distinctive black baldy that was present in neither parental breed but that neatly and clearly stamps the breed heritage behind the crossbred animals.

Another popular crossbred in some areas is the "blue suckler" that results from crossing white Shorthorn bulls with either Angus or Galloway cows (Figure 9.6). A similar color results from crossing white Shorthorns to Holsteins or Friesians. These crosses produce

Figure 9.5 A Hereford to Angus cross consistently produces "black baldy" calves.

well-adapted cows that produce abundant milk for rapidly growing calves. In this instance, the white Shorthorn is providing recessive *red* and dominant *roan*. The Shorthorn's whiteness assures homozygosity for the *roan* allele. The black parent provides *dominant black* and recessive *wild* (non-roan). The resulting calves are therefore all *roan* and *dominant black*, which gives a blue roan that is distinctive, unlike either parental breed, and clearly identifies these as the deliberate crossbreds they are.

Similar strategies produce grey (dun) calves from mating Charolais cattle to a black breed (Angus or Galloway). Some breeders prefer these, others shun them because of the

Figure 9.6 Blue roan cows are generally highly regarded due to the specific breeds crossed to produce them.

occasional risk of a "rat tail" calf that will be discounted in some markets. Some markets discount all grey duns for this reason, even though the duns are much more common than the true rat-tail calves. If the goal is to produce the grey dun color, then mating a very white Charolais to any black breed will accomplish this. The Charolais provides *charolais dilution* and *red*, the black parent provides *wild* (non-dilute) and *dominant black*. Charolais cattle can also provide *pinzgauer* or other spotting alleles in some cases.

A similar strategy also provides for the production of yellow cattle with white faces. These result from crossing Charolais and Hereford mates. These breeds both have recessive *red*, so color variation at *Extension* does not result. However, the parental breeds still leave evidence in the crossbreds that receive dominant *charolais dilution* from one parent, and the *hereford* white face from the other. The resulting crossbred calves recombine the parental alleles into a distinctive phenotype unlike either, but showing evidence of both.

Strategies can also be used to avoid certain combinations. If dun cattle are not desired, usually due to avoidance of rat-tailed calves, then breeders have a few options. If it is still desirable to use a Charolais parent, then matings to black breeds are problematic. However, some Charolais cattle are heterozygous for *charolais dilution*. These are usually more yellow than the clean white of the breed. That yellower color of Charolais can be used with black mates to decrease the incidence of dun calves, although it only takes that incidence down to 50%, not to zero. A second route is to use Simmental cattle that lack *simmental dilution*. Within the Simmental breed many cattle lack the *simmental dilution* and therefore cannot pass it along to their crossbred calves.

A final example of specific trademarked crossbreds is the generally uniform brindle white-faced calves that result from mating Hereford cattle to American Brahman cattle. This result can be a bit of a surprise because the brindle color is not present in either parent breed. The Hereford in this example contributes the dominant *brindle* allele at the *Agouti* locus, even though it is masked by the recessive epistatic *red* allele at the *Extension* locus in this breed. The American Brahman parent provides the dominant *wild* at *Extension*, which then unlocks the expression of the brindle color in the crossbred calves. This is a good example of the complexity of matching up the various recessive and dominant alleles in each parental breed, because the dominant contribution of the American Brahman is not all that obvious in the results even though it is essential to that final distinctive phenotype.

CHAPTER 10

Alpaca and Llama Color

Llamas and alpacas are closely related species whose homeland is South America. The two species share a wide array of colors, but each of them also has a few colors that currently are limited to only that one species and not shared by both. The hair coat of the two species includes a wide range of types. Llamas vary from having relatively coarse short hair over the entire body, all the way to long, uniformly fine fleeces over the entire body with short hair only on the head, belly, and lower legs. Alpacas have been bred specifically for fiber production, and as a result they all have fine fleeces that cover the body and neck, but with short hair on the head, belly, and lower legs.

Fleece types in both alpacas and llamas include two types. Huacaya fiber is the more common fleece type. It is similar to the wool of Corriedale sheep with a blocky staple structure, definite crimp, and fine fiber diameter. Suri fiber is long, silky, and lustrous. It usually has a pencil lock structure and minimal crimp. Suri fiber is superficially similar to mohair or luster longwool wool, but is distinct enough to be classified as its own unique fiber type. The long fibers of both of these highly developed fleece types have a very high potential to hold dark, intense colors of both pigment types. This is unlike the fibers of sheep and goats. The potential for fully saturated pigment is equally true of coarse primary fibers and finer secondary fibers, which is another contrast to the other species.

North American alpaca breeders have long touted the fact that alpaca fiber comes in 22 natural colors. While this is true, these colors occur along a continuum so the actual number of shades is nearly infinite. Teasing out the details of the genetic control behind this variation can be important if breeders desire a certain range of colors. Some superficially similar final colors, such as grey, can result from very different mechanisms that include either mixing white and colored fibers, or diluting all fibers to a uniform blue-grey color. Though the final yarn color from these two mechanisms is similar if processed by commercial mills, they can each yield noticeably different results following the hand-processing of a single individual fleece.

Llamas descend mostly from the guanaco, and alpacas descend mostly from the vicuña. Each of the domesticated species also has minor contributions from the other wild species. Both the guanaco and vicuña still exist in the wild today and they offer a glimpse into the

Figure 10.1 The guanaco is the ancestor of the llama. Its tan body is paler ventrally, and the head is dark grey or nearly black.

original color from which the colors of llamas and alpacas are derived. Both wild species have tan pheomelanic bodies with dramatically lighter ventral regions leading to a distinctly shaded pattern (Figure 10.1). The tan body is accented with minor black regions on the head and lower legs. Vicuñas are generally paler than guanacos, with light tan pheomelanin and fairly restricted black eumelanic regions. The guanaco has darker tan pheomelanin, more obvious black stripes down the fronts of the lower legs, and a uniformly colored head that is grey to nearly black. Neither of these wild species has the dramatically striped head patterns that are typical of sheep and goats. The ancestral color patterns of the camelid species persist in their domesticated descendants more commonly than the ancestral colors of cattle, sheep, goats, and hogs.

Genetic control of llama and alpaca colors has not been studied as thoroughly as has that of other species. Many studies are difficult to decipher because they concentrate only on characterizing the color of the fleeced portions of the body. This can easily lead to confusion because several distinct color patterns have similar body colors, and when color is only classified by fleece color these can all be grouped together in a single class. Many colors can only be distinguished from one another by the color of the shorthaired regions of the head and lower legs. Ignoring the shorthaired regions that lack fleece leads to inadequate color classification, and this in turn leads to misguided conclusions about inheritance. A good first step in studying color is always accurate and repeatable phenotypic classification. This must occur before meaningful genetic results can be obtained.

Recent studies, especially in alpacas, have included segregation data as well as molecular investigations into the DNA sequences behind the different colors. Some studies have used very complex theories, with the unfortunate consequence that nearly any theory can be substantiated by a data set that is carefully selected. Additionally, many of the circulating theories ignore the basic homologies of color inheritance that have proven so useful in unravelling the details of the inheritance of color that are similar across many species. Understanding the similarities across species may seem trivial, but it can be a powerful aid in understanding basic genetic mechanisms, as well as their role in producing various final phenotypes. This chapter attempts to make sense of variation that is seen in these species, and to link the phenotypic variants to results from breeding programs so that the genetic control can be understood.

This chapter focuses mostly on alpacas because this species has undergone more investigations than llamas. Alpacas are used mostly for fiber production, and color is important for the end uses of the fleeces. Production in South America includes many large herds that are managed as commercial enterprises. Alpacas in those regions are also often kept in small herds by subsistence farmers. Those two management environments put different emphasis on selection for color, with many of the large commercial herds heavily favoring white fleeces.

Herds of alpacas outside of South America have generally not had intense selection favoring white fleeces. This somewhat relaxed selection environment provides more opportunities for the expression of a wide array of color variation. The variable colors reflect the complexity of interactions at multiple loci. For many breeders, this variation has led to an unsettling lack of ability to accurately predict color outcomes. However, understanding the basics of the underlying genetic control of color can allow breeders to narrow the range of colors produced in offspring. Over time their experiences can provide for consistently predictable color production. The genetic control of color in llamas and alpacas fortunately is similar to the control in other species, and understanding the basics can greatly assist breeders in reaching their goals.

10.1 Extension *locus*

The *Extension* locus is not a frequent source of variation in the color of llamas and alpacas. Most alpacas and llamas have the *wild* allele at this locus, which yields the control of color to the *Agouti* locus. Even though this is the general rule, some few breeding results are indeed consistent with variation in *Extension* alleles.

Some alpacas have production records that are consistent with the presence of a *dominant black* allele. The evidence for this is that occasionally black animals produce black offspring, as well as offspring with easily recognizable *Agouti* locus patterns, following mating to other black animals. This happens rarely, but is consistent enough to provide evidence for the presence of a *dominant black* allele. This allele, though rare, can easily distort color segregations in herds where it is present.

Alternatively, occasionally two alpacas with obvious *Agouti* locus patterns produce uniformly light tan (fawn) offspring. This is consistent with a recessive *red* allele at *Extension* that results in a completely pheomelanic color. The frequency of this allele across all alpaca populations is not determined, but it leads to some unusual ratios when present in a herd.

These two mutant alleles (*dominant black* and *red*) occur only rarely, but in those populations they are present, they can be the most important factors in color production. Despite their rarity, astute breeders can use them to good advantage by identifying animals with them and designing breeding programs to assure desired results. For example, the *red* allele could quickly result in a uniformly fawn herd that would yield no other colors as surprises. In contrast, *dominant black* would assure frequent production of black offspring, but in most cases several other colors would also be produced because this allele can so easily mask other alleles at both the *Extension* and *Agouti* loci.

10.2 Agouti *locus*

Alleles at the *Agouti* locus account for most of the variation in color of both alpacas and llamas. The color patterns caused by these alleles do not have consistent names, which contrasts to the situation with goats and sheep where each *Agouti* variant tends to be distinct and readily identified by breeders. The following discussion attempts to standardize nomenclature, although it is only a suggestion for classifying the consistent phenotypes of these species.

The general trend of *Agouti* locus patterns is that they are in a series that goes from mostly tan in the more dominant alleles, to mostly black in the more recessive ones. This tendency has been validated by the way these colors segregate out in populations where the various colors occur. The *Agouti* locus patterns are all symmetrical. They can be classified by the relative extent of the black and tan areas. The extent and distribution of black in these camelid species occurs as a continuum from minimally black to maximally black with no distinct boundaries where one pattern stops and the next one begins. This is a similar situation to the one in cattle, and is quite different from the situation in sheep and goats where the patterns are distinct and discontinuous which makes each pattern clearly and easily identifiable. A few general principles still hold true when segregation data are evaluated, and these can be useful to breeders despite the lack of definite and clear divisions between the camelid *Agouti* patterns.

The most dominant *Agouti* allele is *white/tan* and produces animals that are almost entirely tan-based. These are usually completely white or fawn (Figures 10.2 and 10.3), although they can more rarely be a deeper and more red-tinged color (Figure 10.4). Whatever the final shade of color, all of these animals usually have at least some minor black trim around the eyes, mouth, and toenails. These black areas, though minor, are a detail that helps to validate the patterns as resulting from *Agouti* alleles because fawn animals from an *Extension* mechanism lack any ability to form black pigment in hair. White animals with dark eyes often have this allele.

As the alleles proceed down the scale of relative dominance, black areas are sequentially added and become increasingly obvious. The next pattern is tan with minor black trim caused by the *vicuña* allele. These are similar to tan animals, but have more extensive black on the nose, around the eyes, and on the tips of the ears. On the feet the black area can

Figure 10.2 Most white alpacas have *white/tan* at *Agouti*.

extend up the front of the leg a short distance. These black areas are obvious but are not very extensive. The base color of the coat varies from light tan to darker red (Figure 10.5).

The next pattern in the sequence is caused by the *guanaco* allele. The body is red with more extensive and obvious black trim on the head and lower legs. The leg markings are distinctive with obvious black stripes down the fronts of the legs. This is the wild-type allele of the guanaco, although in that species the final pattern also has a few other modifications that provide for the light ventral areas that are dramatic and obvious in the original color. The light areas are likely added by modifiers at other loci. In the *guanaco* pattern the depth of tan tends toward red, rather than the lighter fawn color that is typical of the previous alleles (Figure 10.6).

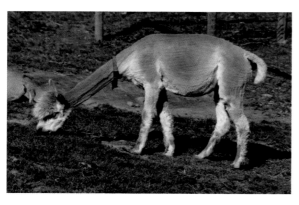

Figure 10.3 Shaded fawn animals are common in alpacas. This color likely results from either the *white/tan* allele, or *vicuña*.

Figure 10.4 Fawn alpacas with minimal shading are usually darker than the ones that shade to lighter colors ventrally.

Animals with the *bay* allele usually have dark red bodies with solid black lower legs and solid black heads (Figure 10.7). The intensity of the shade of red can be quite dark, which is unusual in other fiber-bearing species. As a result, fiber from both of these species offers unusual color options for fiber artists. These dark red colors are simply unavailable in fibers from other species, especially in fine fibers.

Black animals with a tan belly have the *black and tan* allele. This is a rare allele, and also

Figure 10.5 Entirely red alpacas with minimal black trim are fairly rare (photo by E. Kinser).

Figure 10.6 The guanaco type color has obvious black areas.

stands out as different from the other alleles in causing a phenotype that is not along the continuum caused by the other alleles. These animals are nearly entirely black but have pale bellies, and the animals with this pattern therefore yield black fleeces. In contrast to other species with this sort of pattern, the tan is usually limited to the belly and does not result in striped patterns on the lower legs or head. This pattern can be easily overlooked if animals are not closely inspected because in many of them the only tan regions are those with short hair, and these can be effectively hidden in animals with long fleeces.

Animals with the *mahogany* allele are extensively black, but retain dark red or tan areas that have mixtures of tan and black fibers (Figure 10.8). These mixed areas are in the flanks

Figure 10.7 Alpacas with the *bay* allele have distinctly black heads and legs.

Figure 10.8 Alpacas with the *mahogany* allele are nearly black with somewhat lighter ventral areas.

or near the belly. The overall appearance is a black animal dorsally, shading to a more mahogany color ventrally.

Solid black animals have the *nonagouti* allele. This is the most recessive allele at the *Agouti* locus and is the most common cause of black color in these species. Some black animals fade due to sunlight exposure. Many do not fade, and instead maintain a jet-black color that is unusual in other fiber-producing species (Figure 10.9).

Figure 10.9 Black alpacas are noteworthy for minimally fading fleeces, in contrast to the fleeces of most sheep and goats.

Most of the color patterns caused by these alleles, especially *white/tan, vicuña, guanaco, bay, mahogany,* and *nonagouti* are fairly distinctive when the most typical examples are considered. The actual color variation in large populations occurs much more as a continuum rather than as a series of steps with abrupt distinctions between each pattern. This opens up the very real possibility that there are either more (or fewer) alleles than postulated here. The *black and tan* allele stands in contrast to the continuum of these other patterns, and is more clearly identifiable as a distinct allele.

Despite the limitations inherent in completely repeatable color pattern recognition, the fact remains that the results of color production in most herds follow the expectation that more extensively tan patterns are dominant, and more extensively black patterns are recessive. This is typical of the *Agouti* locus. The few exceptions to this rule are all consistent with *Extension* locus phenomena controlling completely black and completely tan phenotypes, although such results are quite rare and do not exist in most herds. Breeders can therefore take advantage of the fact that the general pattern at the *Agouti* locus is that tan areas are dominant, while black areas are recessive. The opposite is true at *Extension*.

10.3 Mosaics of Tan and Black

Occasional animals, usually llamas, have an interplay of tan and black areas that are not symmetrical, and therefore are not likely to be caused by the *Agouti* locus. These areas are often on the head, and the genetics of the pattern is uncertain. These usually lack the striped character of brindle animals, but are similar in having tan and black areas in no repeatable distribution (Figure 10.10).

Figure 10.10
Occasional animals have random distributions of tan and black areas.

10.4 Dilution

Dilute fiber in both llamas and alpacas is reasonably rare, with the exception of the light tan called "fawn" in these species. While this light color could arise through the action of specific diluting mechanisms related to individual alleles, in these species it is more likely that this pale color is inherent in the function of the more extensively tan *Agouti* alleles. Some *Agouti* alleles appear to directly cause the lighter shades of pheomelanin, making the darker shades (toward red) difficult or impossible to achieve with some specific *Agouti* alleles. Fortunately for breeders interested in a full range of pheomelanic colors, the alleles with more extensive black areas on the head and legs do allow fully intense dark red pheomelanin.

The dramatically light ventral areas of the wild species, and of many llamas and alpacas, is likely an independent modifier added onto nearly any tan-based color pattern. In horses this effect is called *pangaré* and is due to a dominant allele at a separate locus. The similar effect in llamas and alpacas is also most likely dominant. Genetic studies based on segregation data have not yet established this conclusion.

Allelic variation at the *Brown* locus has never been documented in alpacas and llamas. This can seem to be a contradiction to many breeders because a great deal of fiber from alpacas is described as "brown." This color in alpaca fleeces is generally a very dark red (as betrayed by black head and legs), or is the result of a mahogany mixture of red and black fibers. In contrast with this general trend, a very few animals in these species do appear to be a uniform brown. This is consistent with the action of *Brown* locus alleles. The results of mating these animals to those of other colors have not been available. This color is rarely accurately identified, and such animals are likely to be considered in with the "brown" fiber that *bay* animals produce. This sort of confusion points to the need for accurate color identification before meaningful genetic studies can be accomplished. The brown animals that have a *brown* genotype have a more uniform color than *bay* animals. Animals based on *brown* are only rarely encountered, and produce a fiber with a flatter, cooler brown color than is typical of the dark pheomelanic colors that are so commonly produced by these species. Identifying and using a recessive *Brown* locus genetic mechanism could provide very distinctive shades of fiber.

A few alpacas have a relatively uniform bluish grey color that is consistent with a dilution of black. These animals have fleeces that are lighter than the shorthaired areas. This steel-blue color is rarely encountered. No breeding results have been available so no theory is possible on the genetic transmission of this color.

Several llamas have a unique silvery dilution that causes a very light fleece (Figure 10.11). The fleece color is silvery-grey on a black background, or very light tan on a red background. These animals have full saturation of pigment in regions with short hair, but are dramatically lighter in regions with longer fleece. The genetic control of this variant is not documented. It appears in some parent and offspring pairs, suggesting a dominant mode of transmission. However, it also can occur following mating of two dark animals, which is consistent with a recessive mode of action. Extensive data sets are not available, so a final determination is not possible at this time. This variant is either rare or absent in alpacas, although the colors produced by the mechanism are distinctive and would have appeal to fiber artists.

Figure 10.11 One form of dilution in llamas leads to a dark head and a silvery body. The llamas in A retain considerable pigment over the body (photo A by Stone Crest Llamas). The llama in B is paler, and likely has a different *Agouti* allele as the background for the final dilution (photo B by WoodsEdge Farm).

10.5 *Harlequin / Appaloosa*

"Appaloosa" patterned llamas and alpacas are different from the patterns that are known by that name in horses. The background color in llamas and alpacas is not white, but is instead a dilute shade of whatever color the darker spots are (Figure 10.12). It is therefore likely that this color is due to some type of dilution rather than a white spotting pattern.

Figure 10.12 Harlequin alpacas have colored spots on a background that is nearly white on young animals and then becomes darker with age (photo by E. Kinser).

Figure 10.13 The final appearance of harlequins depends on the extent of the darkening (photo by E. Kinser).

The genetics of this pattern have never been well established. Some evidence suggests that the genetic cause may well behave differently depending on the specific background color. Some breeders indicate that it tends to be dominant on a black background, while recessive on a tan background. Young animals can be dramatically "white with spots," but the pale background areas then become darker as the animal grows and matures. This becomes more obvious with successive shearings (Figure 10.13).

10.6 White Spotting

Several different patterns of white spotting occur in llamas and alpacas. It is likely that each is under separate genetic control, as is the case with other species. There is no evidence one way or the other as to whether they are alleles at a single locus or at multiple loci.

Piebald spotting is one pattern of white spotting, and is likely due to a *spotted* allele similar to that in other species. Animals expressing *spotted* generally have color remaining around the eyes as eye patches, and usually white encircles the neck at some point (Figure 10.14). White is usually expressed ventrally and on the legs. The inheritance of *spotted* is somewhat confusing. Many llama breeders insist that it is recessive. In some herds, though, it has dominant characteristics and passes reliably to some offspring of solid animals mated to *spotted* animals.

Figure 10.14 Alpacas with the *spotted* pattern usually have white on the legs, head, and neck (photo by E. Kinser).

At the least white end of the range of expression, *spotted* animals can easily be confused with nonspotted animals. At the most extensively white end of the range they can be confused with white animals. This is especially true if color designation only considers the fleeced portions of the animal, because extensively white *spotted* animals usually retain color only on the shorthaired head. In that case, the fleece is entirely white and the animal is likely to be mistakenly misclassified as white. The *spotted* pattern can be superimposed over any background color, but is more obvious on the darker colors because the contrast is greater. This pattern is more common in llamas than in alpacas, but does occur in both, and by itself is no indication of past crossbreeding of the two species.

The *tuxedo* pattern can be confused with *spotted* but is a different pattern under separate genetic control. It is sometimes called "caped" as a translation from the Spanish "capirote." This pattern generally has white on the underline from the head to the rear (Figure 10.15). Most *tuxedo* animals have color remaining on the back of the neck, barrel, and back, and are white on the face, lower neck, legs, and belly. This pattern can be superimposed over any background color. It is probably inherited as a dominant gene. This is superficially similar to the pallor of the head and neck on *grey* animals, but is a distinct pattern that can be combined with any base color. It should not be confused with *grey*, as the two patterns are genetically distinct.

Some llamas and alpacas are white with dark heads and feet (Figure 10.16). This pattern

Figure 10.15 The *tuxedo* pattern of white spotting is ventral.

is almost certainly not one of the other spotting patterns, such as *spotted* or *tuxedo*. The logic for this conclusion is that the *spotted* and *tuxedo* patterns nearly always involve white on the head and legs. Therefore an extensively white animal produced by these mechanisms is expected to have obvious white areas on the face and legs instead of allowing these areas to remain colored on an otherwise white animal. Extensively white animals with this pattern consistently retain pigment on the head and lower legs, which is not expected of the other two patterns. The details of the inheritance of this pattern are not documented. These animals are frequently classed as white when color designation is only based on observations of the fleece.

A distinctive pattern in alpacas is *speckled* (Figure 10.17). The minimally white manifestations of the pattern have distinctive small white spots along the belly and lower neck,

Figure 10.16 Many llamas and alpacas have white bodies with color on the head and lower legs (photo by WoodsEdge Farm).

Figure 10.17 The *speckled* pattern is uncommon in alpacas. The extent of the white is quite variable.

which extend up onto the sides. More extensively marked animals can be nearly white, with colored heads and small to medium sized colored spots scattered throughout the coat. In the minimal and medium ranges of expression this variant results in a very interesting fleece which might superficially resemble grey or harlequin as a shorn product. Details of the inheritance of *speckled* are uncertain. This rare pattern does occur in outbred alpacas, and in those families it occurs in both parents and offspring. This suggests that it is due to a dominant allele.

A few other distinct patterns of white spotting occur in these species but are poorly characterized. Occasionally llamas have large round, colored spots on a white background. It is different from any pattern in alpacas, and is likely its own unique pattern caused by a specific allele. The pattern of inheritance and genetic control has not been documented. Suri alpacas, especially darker fawn suris, frequently have white areas in a distribution not expected of the other patterns. The minimal contrast between white and fawn makes characterizing the exact pattern difficult.

Markings on the head can be especially difficult to characterize. Large patches of white on the head are likely to be betraying the presence of a minimally expressed spotting pattern (*spotted* or *tuxedo*). This can be critically important if certain colors or patterns are either favored or not favored in the progeny. A similar line of thinking holds true for most large white patches involving the feet and lower legs. At the other extreme are minimal white marks on the head, which are reasonably common on the heads and legs of alpacas and llamas. They usually do not affect the fleeced portions of the animals, and so are usually overlooked as being important to color identification or to genetics. Frequently these are small enough to be inconsequential in fiber-producing herds. The genetics of these is probably complicated, as is the case in most other species.

Odd, small white patches occasionally occur along the belly or sides. These are probably not related to major spotting genes because their distribution does not fit well with any of those. Animals with these spots can probably be used in a breeding program much as any truly solid-colored animal. Such spots can occur in fleeced portions of the body, and can therefore affect the fiber shorn from these areas.

10.6.1 Grey

Grey phenotypes are fairly common in alpacas, but rare or non-existent in llamas. Alpacas with the *grey* pattern generally have white, or nearly white, heads and legs. The coat over the ears and top of the head is usually dark. The body coat is a combination of discrete areas with dilute (grey), dark, and intermixed hairs (Figure 10.18). The result is usually a fleece with some variation in color from region to region. These animals betray their peculiar mixture of areas with varying colors when they are freshly shorn because the interplay of dilute, dark, and mixed areas becomes more obvious (Figure 10.19). This pattern somewhat resembles the appearance of the *pedi* pattern of Nguni cattle. Dark *grey* animals may be confused with nongreys, and light grey animals may be confused with white animals.

The *grey* pattern is due to a dominant allele. However, the darkest ones do not behave genetically the same as the medium and light ones and therefore may be due to a different genetic mechanism. Dark grey alpacas tend to reproduce their grey color more rarely than the medium and light grey animals. Some of this tendency may be due to very dark *greys* being misclassed as nongreys, which upsets segregation data due to incorrect classification of the offspring. Equally, the darkest greys may have a cause other than the *grey* allele.

No homozygous *grey* animal has ever been documented. The proof for this would be a *grey* that only produced *grey* and no nongrey offspring following mating to nongrey mates. In every instance where a *grey* animal has been mated to nongreys in sufficient numbers, some nongrey offspring have been produced. This demonstrates that the *grey* parent did

Figure 10.18 The final appearance of *grey* alpacas depends on the background color (photo by E. Kinser).

Figure 10.19 Alpacas with the *grey* allele nearly always have dark spots.

not have two copies of the *grey* allele. In addition, *grey* to *grey* matings consistently produce about two *grey* offspring to every nongrey. This also points to a lack of homozygous *grey* animals. This ratio is typical of a gene that is lethal to homozygotes. In this example the homozygotes are likely lost very early in gestation, so no overt loss is noticed.

The nomenclature of *grey* alpacas is important to breeders. On a black background, *grey* yields silver colors (dark, medium, or light). On a mahogany background or a dark bay background the result is a brown type of rose *grey*. On a bay background the result is a redder version of rose *grey*. On a fawn background *grey* can be difficult to appreciate, and is likely to be overlooked as contributing to the final color. The tip-off to the presence of *grey* on these pale colors can be the pale head and legs, although this phenotype can easily be confused with white-faced fawns that lack *grey* as the cause for their color.

The dark spots on *grey* animals are usually the background color. On silver *grey* these are black, on brown rose *grey* they are brown, and on red rose *grey* they are red. Many *grey* alpacas, though, have dark spots that vary considerably. Silver *grey* animals often have brown or red dark spots in addition to the expected black ones.

10.6.2 Roan

The *roan* pattern is rare in North America. The bodies of these animals vary from red roan to black roan depending on the base color. These animals have dark heads and legs, with the color of these regions depending on the base color of the animal (Figure 10.20). This pattern behaves as a dominant allele, much as is the case with similar patterns in other

Figure 10.20 The *roan* pattern lacks the white face of *grey* alpacas.

species. Animals with *roan* are very distinctive, although fleeces are similar to the more usual *grey* that has a white head. The dark-headed *roan* pattern can be superimposed over any background color, and is especially stunning on a red or black background. The *roan* pattern lacks the distinctive random dark spots of *grey*, and instead has a very uniform mixture of white and colored hairs throughout.

Some roan phenotypes do not have uniform roaning, but instead have a heavier presence of white hairs on the lower neck and lower sides than they do up over the back. This variant could be easily overlooked and therefore is prone to being classified inaccurately. These shaded *roan* animals appear in the same families as dark-headed *roan* and are probably due to the same genetic mechanism. Shaded *roan* occurs most commonly in suris and more rarely in huacayas. Shaded *roan* animals are often only minimally roan, and the minor roan areas are usually restricted to the sides.

10.6.3 White

White alpacas and llamas are a confusing group. Many different genetic pathways can produce a white animal. The final results are similar and are therefore easily confused as to which of the different mechanisms is working. Common strategies for arriving at a white animal include selecting for the most dominant *Agouti* locus allele, *white/tan*. This allele likely expresses more often as white than as pale fawn. A second common strategy for producing white animals is to select for those that are extensively white spotted. These animals are essentially one big white spot. Many of them retain a few pigmented areas on the head or lower legs.

Another strategy that is fairly commonly used to produce white alpacas is based on combinations of the *grey* allele with one of various spotting alleles. The result is a blue-eyed white animal with a starkly white fleece. These animals are deaf, and therefore do raise concerns about animal welfare issues. Most people who have such animals note that they fit well into the social structure of their herd and are not consistently at the bottom as would be true if they were compromised in their ability to discern social cues. If intelligently and humanely managed these animals do quite well. One advantage to these animals is the starkly white fiber. One disadvantage, aside from their deafness, comes from the specific genetic combination that produces them. These white animals can always produce a wide array of colors in their offspring, including solid colored, *grey*, *spotted*, or *tuxedo*, or animals with minor white marks. When these blue-eyed white animals are used with solid-colored mates, several of these patterns can emerge alongside a minority of the desired white offspring.

Breeders interested in avoiding the production of blue-eyed white alpacas do need to be careful when mating *grey* animals. Alpacas with *grey* have the potential to produce blue-eyed whites following mating with any white spotting pattern, and this includes even relatively minor white marks on head or legs. Breeders interested in avoiding the production of blue-eyed whites should mate *grey* animals only to solid-colored animals that lack white marks.

CHAPTER 11

Putting Knowledge to Work:
Alpacas and Llamas

Alpacas and llamas share a great many alleles that lead to the wide array of colors in these species. It is generally true that alpaca breeders are more focused on color than llama breeders are, for the simple reason that fiber is of more direct consequence to alpaca breeders. Color is important to some llama breeders, especially those with Argentine llamas and their fine fleeces, or those with the suri fleece type. Both of those types of llamas tend to be raised specifically for fiber production, and color variation is important in meeting demand.

Prediction of color outcomes in the mating of alpacas is fairly straightforward in most populations because color is most often controlled only by the *Agouti* locus. Where this is true, the more extensively tan colors are dominant to the more extensively black colors. This opens up opportunities for breeders interested in producing a full range of colors in offspring. This is possible even if they are working with only a small number of animals (Figure 11.1). If a wide range of color production is the goal, then a strategy for success is

Figure 11.1 Many alpaca breeders deliberately favor color variation and successfully manage it even in small herds (photo by E. Kinser).

to use a recessive black male over the entire range of colors. A potential detraction of this strategy is that over several generations the relative proportion of black animals produced will increase because the females will all be heterozygous for *nonagouti*. However, as long as breeding females are selected to cover the entire range of colors, then around half of the offspring will include those other colors and these can be selectively retained in the herd. This assures a balance of colors, even if it also assures an excess production of black animals to be sold to others.

The genetic status of the black male as recessive is important, because alpacas have *dominant black* at *Extension* as a rare option for producing black animals. One way to assure that a black animal has *nonagouti* at *Agouti*, rather than *dominant black* at *Extension*, is to locate black animals that have parents with other obvious *Agouti* locus colors. Colors of parents that are not black and instead are *Agouti* patterns assure that the black offspring cannot have the *dominant black* allele. This strategy is also useful for those breeders interested in a uniformly black herd that produces no other colors as surprises, because it assures that all animals are *nonagouti* at *Agouti*.

An opposite extreme is to select preferentially for white animals. Unfortunately, several different genetic pathways lead to white in alpacas and not all of them can be distinguished accurately one from another solely on the basis of phenotypic evaluation. The consequence of this is that a herd of white animals assembled from a wide variety of sources is very likely to include several different genetic mechanisms. That can make the accurate prediction of color outcomes difficult or impossible.

Selecting for white alpacas has a few inherent challenges. The most reliable genetic mechanism for the 100% production of any desired phenotype is to use a recessive genetic mechanism. Unfortunately, this is not an option for white alpacas and llamas because most genetic mechanisms for white animals have dominant tendencies. Colored offspring at some frequency or another are therefore likely in most situations, although generally these will only be a minority of the offspring produced as long as all parents are white. The relative proportion of these colored offspring can be important to producers in some situations, but little can be done to eliminate them completely. One strategy to decrease the production of colored offspring is to only use breeding animals produced from two white parents. A step up from that criterion would be to also remove any animals that produce colored offspring. This might prove to be too severe a selection criterion by virtue of removing too many animals, which then reduces the genetic variation needed for viability and avoidance of inbreeding depression.

The most generally useful genetic mechanism for white alpacas uses the dominant *white/ tan* allele at the *Agouti* locus. This allele does not appear to come with any disadvantages. Another strategy for producing white animals is to use *grey* animals mated to spotted ones. This should produce about 25% of the offspring with starkly white fiber and blue eyes. One disadvantage of this strategy is the fact that these white animals are deaf. A second disadvantage is that the specific genetic combination is guaranteed to produce a very high proportion of colored offspring rather than the desired white ones. This genetic mechanism does produce starkly white animals, but can never succeed in producing a consistent majority of them in any population.

From time to time rare, novel colors or white spotting patterns appear in alpaca and

Figure 12.1 The wild-type color of mature hogs is a grizzled mixture of black and tan (photo by J. Beranger).

Several alleles lead to a final black phenotype (Figure 12.2). The specific mutational change of these alleles varies from breed to breed. Even though there are numerous individual alleles, these can all be conveniently grouped together as *dominant black* based on their behavior in breeding programs. The specific mutations are notably different depending on their origin in either Asian or European breeds. Hogs with any of the *dominant black* alleles are born black and stay black throughout their lives.

The *wild* allele is recessive to *dominant black*. This allele allows expression of the *Agouti* locus. The piglets that have colors controlled by the *Agouti* locus have the longitudinal pale stripes characteristic of wild hogs. They lose the stripes as they grow, eventually attaining

Figure 12.2 Several different *Extension* locus alleles lead to a black phenotype. They have been used to establish breed identity in the Large Black and other breeds (photo by J. Beranger).

Figure 12.3 This Mangalitsa piglet is beginning to lose its juvenile stripes (photo by J. Beranger).

one of several final adult colors. Those final colors often change the background color against which the juvenile stripes are expressed, but any piglet with longitudinal stripes has this allele (Figure 12.3).

Recessive to *wild* is the *dark ginger* (E^p) allele. On an otherwise unmodified genetic background the final phenotype is some shade of tan but with round black spots scattered throughout the coat (Figure 12.4). The shade of tan is under independent control. The final shade is usually consistent throughout a breed which provides for breed recognition. The Oxford Sandy and Black, as the name suggests, has a sandy to more red background

Figure 12.4 These Ossabaw hogs both have the *dark ginger* allele, but one has a diluted tan background that is almost white.

Figure 12.5 The final color of red hogs with the *red* allele varies from a light red typical of the Tamworth in A to a much darker red typical of the Duroc in B (photo A by J. Beranger).

color over which are superimposed a varying number of black spots. The Gloucestershire Old Spot has a cream to white background highlighted by black spots. In the case of the Gloucestershire Old Spot at least some tendency for a fairly low number of black spots is present, suggesting that the relative number of the black spots is under independent genetic control and is therefore subject to selection pressure.

The most recessive allele at *Extension* is *red* (E^e). This produces a final phenotype that is entirely tan or red. The depth of the red varies from a medium red shade in the Tamworth, to a much darker red typical of the Duroc and some Spanish breeds (Figure 12.5).

12.2 Agouti *Locus*

The *Agouti* locus is expressed in genotypes that have the *wild* allele at *Extension*. The result is piglets with pale longitudinal stripes against a generally darker background color. A few different adult colors can emerge from this striped beginning. Striped piglets are unusual in most breeds, but the Mangalitsa retains these striped piglets in all three of the colors of the breed: swallow bellied (*black and tan*), red, and blond. How all of these relate to the underlying genotype at all loci is problematic because some piglets of the breed lack stripes

Figure 12.6 The *black and tan* color of Mangalitsa hogs is called "swallow belly" (photo by J. Beranger).

but still achieve these adult colors. The usual cause of red in most hog breeds, for example, is E^e. This genotype lacks striped piglets. Against that background of some uncertainty, only three alleles have been documented at the *Agouti* locus of hogs.

The original *wild* allele at *Agouti* results in a final wild-type coat color that is made of hairs that are banded black and tan. This gives the grizzled appearance typical of mature wild hogs. The banding pattern on the hairs is fairly uniform throughout the coat, in contrast to the more regionally distinct patterns that are typical of the other species with their interplay of black and tan areas. The wild-type color is exceedingly rare in domesticated hog breeds.

The *black and tan* allele produces a typical pattern of black with a tan belly that extends to the inner and lower legs as well as the bottomline of the neck and the lower portion of the head (Figure 12.6). There are small spots above the eyes, and the lower portion of the ears is tan. This pattern is fairly rare or non-existent in most breeds. It occurs in the Mangalitsa breed along with several other colors.

The most recessive allele at *Agouti* is *nonagouti* (A^a) which results in a completely black phenotype. This genetic mechanism is used only rarely to produce black hogs, with *dominant black* being much more common. This black that results from *nonagouti* cannot be distinguished visually from *dominant black*, although each behaves very differently in a breeding program.

12.3 Dilution

One of the most common dilute phenotypes in hogs reduces tan pigment. The consequence for *dark ginger* hogs is to achieve a phenotype of black spots on a cream or white background instead of the obviously tan or red background usually associated with the *dark ginger* allele (Figure 12.4). Many different genetic mechanisms seem capable of diluting pheomelanin in this manner, and unfortunately none is very well documented as to the precise mode of genetic action. Some of these mechanisms are dominant, including one that is described as "chinchilla" but whose locus of action is not defined. In most other species *chinchilla* is at the *Albino* locus (*Tyr* or *Tyrosinase* locus of the mouse). In nearly all other species this allele is recessive, raising doubts that the hog allele is homologous with these others even though the final phenotype is similar.

Another specific genetic mechanism is characterized as *dilute*. This modification lightens tan from redder shades to paler shades. This action is only moderate, and does not yield a dramatically light final coat color. Instead, the color remains identifiable as tan rather than as white.

At the *Brown* locus (*TYRP1* for *Tyrosinase Related Protein 1*) a recessive *brown* allele lightens black to blond. This is a dominant allele and is present in some Chinese breeds. It is unlikely to be the same mutation that leads to blond Mangalitsa hogs, which is a very pale cream color.

A rare allele at the *Pink Eye* locus is recessive *pink eye*. This allele lightens black to a sepia color, and also causes the eyes to be pink rather than the usual brown of hog eyes. This allele has not been used to standardize any breeds.

12.4 White and Related Colors

Many breeds of hogs are white because the skin of the final meat products lacks pigment which tends to be favored in many settings. Several different genetic mechanisms produce white color. Each of them behaves differently in a breeding program. They each tend to be uniform throughout the breed in which they occur.

The *Spotting* locus in hogs more often goes by the name *KIT* which is adopted from the mouse nomenclature. This locus is a common site for alleles that cause white spotting in animals, and hogs are no exception. The action of mutant alleles at this locus in hogs is quite variable and depends on the specific allele. In many species, including hogs, the *Spotting* locus hosts a large number of alleles. Many of these alleles produce phenotypes that overlap somewhat. A completely accurate one-to-one relationship of phenotype to genotype is therefore difficult to achieve. Most of the alleles at this locus result in a largely white phenotype, especially in homozygotes. These are epistatic to other loci, and mask their effects in the phenotype. These mutations are all dominant to the *wild* allele at this locus. Hogs with the *wild* genotype at this locus express color as determined by the other loci.

The most dominant allele at *Spotting* is *lethal white*. The "lethal" part of the action is only present in homozygotes. Heterozygotes are perfectly viable and functional, and are completely white. This allele is not used very frequently to produce white hogs, but does

Figure 12.7 This Blanc de L'Ouest hog has the white color typical of mutants at the *Spotting* (*KIT*) locus (photo by J. Beranger).

account for the white color of a few breeds. Those breeds routinely produce colored piglets, in addition to the more desired white ones, because the white hogs are all heterozygous. This allele has the advantage of producing white piglets, but a significant disadvantage is the loss of about 25% of embryos. Litters that are produced from mates that are both heterozygous for this allele therefore have smaller numbers of piglets than would be the case if they were mated to animals without this allele. In addition, most litters where both parents have this allele have about two-thirds white piglets, and one third colored piglets due to the consequences of the heterozygosity of the parents as well as the lethality of the allele to homozygous embryos.

At least three different alleles are grouped under the single designation of *dominant white* (Figure 12.7). These are all dominant alleles that result in a completely white phenotype. White hogs with any of these alleles are fully viable as homozygotes. Many white breeds use these alleles as their genetic basis and through that can achieve uniformly white populations with few or no colored piglets ever produced. Due to the lack of lethality, litter sizes in these breeds do not experience losses from the mechanism that causes subfertility with the *lethal white* allele.

Other alleles at this locus have less extreme action (Figure 12.8). The *patch* allele results in a mostly white hog with some residual colored areas along the back. The *roan* allele results in a mixture of white and colored hairs that is more extensive than *patch* phenotypes, and may or may not reside at this locus.

One fairly common phenotype in hogs is belted, which in most breeds is caused by the *belt* allele that is usually assigned to the *Spotting* locus (Figure 12.9). This is the common

Figure 12.8 These crossbred hogs (A and B) illustrate the effects of *patch* and *roan* (photos by D. Newcom).

belt in European breeds, and it varies in expression from narrow and incomplete to covering nearly the entire middle of the hog.

At least one other locus provides a genetic mechanism for white phenotypes. It is active in some Chinese breeds. This relates back to the general phenomenon that similar phenotypes arising in different geographic regions or populations may well have different genetic causes. This type of white is recessive, unlike the *dominant white* at the *Spotting* locus. The recessive white phenotype fully masks *dominant black* and is therefore due to an epistatic allele.

Figure 12.9 The *belt* allele is part of the breed package of the Hampshire hog (photo by K. Kerns).

12.5 White Spotting

Various white spotting patterns occur in hogs. One potential source of confusion related to white spotting in hogs is the similarity of some of these legitimately spotted phenotypes with those achieved by *dark ginger*. The *dark ginger* phenotypes can easily be confused with truly white spotting on hogs where the tan background color is diluted to cream or white, because the final result is a white hog with black spots.

Common white spots, such as seen in the Spotted breed, are recessive phenotypes from the *spotted* allele (Figure 12.10). This is at a different locus from *Spotting*. This pattern is widely distributed in several breeds. This is similar to the phenotype of piebald spotting in other species. The minimal extent of spotting is usually on the head and legs, and then it progressively increases to include body regions. While not an absolute rule, the residual colored areas of hogs with this sort of spotting are usually not the consistently round areas of the *dark ginger* allele and this can help observers to separate out the two types.

There is a distinct allele at the *EDNRB* (*Endothelin Receptor Type B*) locus that has an allele that causes white belts. This allele is present in some Asian breeds. The final phenotype is similar to the *belted* allele at *Spotting*, but the locus of action is completely different. This highlights the complexity of hog genetics, with many phenotypes that mimic one another but that each have completely different genetic machinery.

A "half colored" phenotype is inherited as a recessive and has white on the anterior portion of the hog. This is likely the same allele that is responsible for the characteristic markings of the Hereford hog breed (Figure 12.11). This breed is deliberately selected to have a color and spotting pattern similar to Hereford cattle. Selection of the right extent of white is essential to achieving the final balance of color and white. It has sometimes been called the "white face" allele.

Some hogs have a distinct pattern of minor white on the feet, tail, and nose. These are called "six points white" in Chinese breeds. The genetic basis for this color phenotype is

Figure 12.10 This hog has the *spotted* allele (photo by J. Beranger).

Figure 12.11 The Hereford hog is selected for a specific distribution of white areas over a red background (photo by J. Beranger).

complicated, and involves at least three loci. A consequence of the complex genetic interactions is that the final phenotype sometimes passes along as a dominant trait, and sometimes as a recessive trait. This is consistent with many other species where relatively minor expressions of white occur on the head, feet, and tail, and can be due to multiple genetic pathways all achieving a similar end phenotype.

A unique type of roan or grey phenotype is seen in some breeds including the American Guinea hog and the Choctaw hog. The addition of white hairs over a black background color results in a blue-grey color, and in most breeds these are called "blue" (Figure 12.12).

Figure 12.12 This blue Choctaw gilt has a roan pattern that can be produced by at least two different genetic mechanisms (photo by J. Beranger).

The exact mechanisms behind all occurrences of this color are uncertain. The *Spotting* locus has a dominant *roan* allele that can produce this phenotype. This allele is likely to be the cause of blue piglets in most pig breeds in which the blue color reliably passes from generation to generation as long as blue to black matings are accomplished. This consistent pattern is seen in both the American Guinea Hog and the Choctaw Hog.

Occasionally though, blue piglets are produced in these two breeds following the mating of two black hogs. This is consistent with a recessive mode of action, and therefore is not consistent with the *roan* allele. A different allele at an unspecified locus is acting to produce the blue phenotype in such cases. The underlying genetics of this recessive mode have never been fully investigated.

CHAPTER 13

Putting Knowledge to Work: Hogs

Hog color is rarely a key selection criterion for most producers, but this is changing somewhat with a few breeds that have smaller size. Small breeds are seeing increasing use as companion animals, and breeders often favor unusual colors or combinations. Breeders of local breeds that are used for production also tend to be interested in the color variation that occurs in these breeds, with the result that novel colors or new combinations find a good level of demand. A few examples in different breeds serve as examples of what is possible.

Hogs from the KuneKune breed from New Zealand have a small body size and fit well into the companion animal framework. This breed includes several different colors, and the relationships between them follow the normal rules for hog color inheritance. Colors include ginger (*red*), black (*dominant black*), ginger with black spots (*dark ginger*), and the wild-type grizzled color that is called "brown" in this breed. Any of these colors can combine with white spots on feet, tail, and head. Belts also occasionally occur. Dilution of red to cream also occurs, and can result in cream or white with black spots. Breeders of KuneKune hogs have begun to manipulate the various alleles available to them in order to produce new and different combinations. For example, a flashy pattern of white, black, and red areas comes from combining *spotted* with *dark ginger*. The white spotting phenotype and the "cream with black spots" phenotype that comes from diluting *dark ginger* (with black spots) both have a similar final appearance of white with black spots. The two similar appearances can be difficult to accurately define and can produce unexpected results when mated together.

The American Guinea hog, a unique local American breed, has very limited color variation when compared to the KuneKune. American Guinea hogs are small homestead hogs that are seeing a resurgence of interest for local pork production. Nearly all American Guinea hogs today are black, but historic references also include blue and red hogs. Traditionalists from the original homeland of the breed respond to queries about their childhood experiences with reminiscences of hogs that were very consistent for both conformation and color. The interesting detail is that some of them recount the hogs all being black, others recount them all being red, and yet others recount them all being blue. The present situation is a nearly all-black breed, but this reflects a recent population crash and recovery during which some variation in color may have been lost. In addition to these long-standing traditional colors, a few American Guinea hogs are born with minor white spotting on the nose and feet.

Breeders of American Guinea hogs have a delicate path to tread as they try to conserve this local breed. Defining what is a typical American Guinea hog can unwittingly omit some hogs of the breed that just happen to have a variant that is now rare in the breed. This can include the red and blue colors. At the same time, resurging interest among small scale producers can easily result in fads that emphasize rare colors, or other traits, simply due to their rarity. That approach runs a real risk of changing the breed from its traditional form, especially if those rare traits have come in through introgression of genetic material of other breeds.

White marks on American Guinea hogs can be especially perplexing. Low grades of spotting are probably not due to a single gene, but are instead caused by complicated genetic mechanisms at several loci. They occur from time to time in purebred American Guinea hogs, and they raise the question of how best to manage them. If these hogs with minor white spots are selectively bred to one another it is easily possible to increase the extent of white spotting. The eventual endpoint could easily include hogs with obvious spotting on the body. This can occur even within a purebred population free of genetic introgression. However, it is contrary to the traditional form of this hog breed. In American Guinea hogs the minor white spotting, by itself, is not evidence of past crossbreeding, but putting it into a traditional context is important. In order to manage the breed's phenotype to be close to the traditional solid-colored hog, it is likely best to not mate spotted hogs to other spotted hogs. Instead, they should always be mated to solid-colored hogs that lack any white marks. This strategy reduces the chance of combining high numbers of modifiers that facilitate the extent of the white marks, and that would likely lead to extensive white spotting.

Blue hogs were once common in some bloodlines of American Guinea hog. This color is now rare and approaching extinction. The same is true of the red color that was once common in the breed but which is now rare or extinct. Most of the few occurrences of red or blue piglets in the current breed usually become black at maturity, leading to questions as to exactly which genes are present. Using any blue or red hog preferentially in the breed will increase the rate at which these colors are produced, regardless of the exact mode of inheritance. In the case of dominant alleles, the pathway to increased numbers is rapid and simple. Recessive alleles, in contrast, demand more attention to detailed mating plans that can locate carriers and mate them together.

In the local Choctaw hog breed nearly all of the hogs are black. A few blue hogs persist in the breed, and these include some that are a fairly light blue roan color. The mode of inheritance of this color is uncertain. In addition, occasional Choctaw hogs are the "tan with black spots" color that is encountered in many breeds, and comes from the *dark ginger* allele. Choctaw hogs with this color usually have *pale tan* so that the final color is cream or nearly white with black spots. Both blue and tan with black spots could be increased in the Choctaw hog breed by using breeding animals of these colors with appropriate mates. The black color of most hogs comes from *dominant black*, so the *dark ginger* phenotypes with black spotting on a pale tan background will likely disappear in the first generation only to reappear in future generations after mating carriers together. The mode of inheritance of the blue color is undetermined, but a few litters produced from blue hogs mated to black hogs will quickly establish whether the allele causing the color is dominant or recessive.

In most modern breeds of hogs the colors depend on *Extension* alleles other than *wild*. This has the consequence that piglets of these breeds lack the pale horizontal stripes associated with that allele. Some populations, especially breeds raised in extensive free-range settings, do occasionally produce striped piglets even where there traditionally were none. These piglets, in nearly all cases, are a good indication that some other breed or strain has been introduced into the genetic stock. Throughout most of the USA the main culprit is feral hogs with influence from introduced wild boars. When striped piglets occur in these situations the parents should be culled from production, especially if the goal is to conserve a traditional breed. This is in contrast to populations that are not purebred. In that situation, these striped piglets can safely be used for reproduction in order to enhance color variation.

The Mangalitsa breed from Eastern Europe is a relatively recent import in the USA and Western Europe. This breed has an unusual long woolly coat of hair, and also has an array of colors that are simply not available to breeders of most other breeds in the USA. The Mangalitsa colors are based on the *wild* allele at *Extension*, and most piglets are born with the stripes (Figure 13.1). The adult colors include red, blond, and a color resembling black and tan that is called "swallow belly" because of its similarity to the color of a barn swallow bird. Breeders mating these colors together may eventually produce yet more color variation from reassorting the alleles that cause these three basic colors.

Figure 13.1 This Mangalitsa litter's dam is a blond sow, and the litter contains a wide variety of colors. Several of the piglets have a clear striping pattern. The pattern on a few is more subtle, but still present, and indicates that the colors are not produced by *Extension* locus alleles (photo by D. Kauffman).

Index